T0220577

Biosystems & Biorobotics

Volume 20

Aims & Scope

Biosystems & Biorobotics publishes the latest research developments in three main areas: 1) understanding biological systems from a bioengineering point of view, i.e. the study of biosystems by exploiting engineering methods and tools to unveil their functioning principles and unrivalled performance; 2) design and development of biologically inspired machines and systems to be used for different purposes and in a variety of application contexts. The series welcomes contributions on novel design approaches, methods and tools as well as case studies on specific bioinspired systems; 3) design and developments of nano-, micro-, macrodevices and systems for biomedical applications, i.e. technologies that can improve modern healthcare and welfare by enabling novel solutions for prevention, diagnosis, surgery, prosthetics, rehabilitation and independent living.

On one side, the series focuses on recent methods and technologies which allow multiscale, multi-physics, high-resolution analysis and modeling of biological systems. A special emphasis on this side is given to the use of mechatronic and robotic systems as a tool for basic research in biology. On the other side, the series authoritatively reports on current theoretical and experimental challenges and developments related to the "biomechatronic" design of novel biorobotic machines. A special emphasis on this side is given to human-machine interaction and interfacing, and also to the ethical and social implications of this emerging research area, as key challenges for the acceptability and sustainability of biorobotics technology.

The main target of the series are engineers interested in biology and medicine, and specifically bioengineers and biaroboticists. Volume published in the series comprise monographs, edited volumes, lecture notes, as well as selected conference proceedings and PhD theses. The series also publishes books purposely devoted to support education in bioengineering, biomedical engineering, biomechatronics and biorobotics at graduate and post-graduate levels.

About the Cover

The cover of the book series Biosystems & Biorobotics features a robotic hand prosthesis. This looks like a natural hand and is ready to be implanted on a human amputee to help them recover their physical capabilities. This picture was chosen to represent a variety of concepts and disciplines: from the understanding of biological systems to biomechatronics, bioinspiration and biomimetics; and from the concept of human-robot and human-machine interaction to the use of robots and, more generally, of engineering techniques for biological research and in healthcare. The picture also points to the social impact of bioengineering research and to its potential for improving human health and the quality of life of all individuals, including those with special needs. The picture was taken during the LIFEHAND experimental trials run at Università Campus Bio-Medico of Rome (Italy) in 2008. The LIFEHAND project tested the ability of an amputee patient to control the Cyberhand, a robotic prosthesis developed at Scuola Superiore Sant' Anna in Pisa (Italy), using the tf-LIFE electrodes developed at the Fraunhofer Institute for Biomedical Engineering (IBMT, Germany), which were implanted in the patient's arm. The implanted tf-LIFE electrodes were shown to enable bidirectional communication (from brain to hand and vice versa) between the brain and the Cyberhand. As a result, the patient was able to control complex movements of the prosthesis, while receiving sensory feedback in the form of direct neurostimulation. For more information please visit http://www.biorobotics.it or contact the Series Editor.

More information about this series at http://www.springer.com/series/10421

Maria Chiara Carrozza

The Robot and Us

An 'Antidisciplinary' Perspective
on the Scientific and Social Impacts
of Robotics

 Springer

Maria Chiara Carrozza
Scuola Superiore Sant'Anna
The BioRobotics Institute
Pisa, Italy

ISSN 2195-3562 ISSN 2195-3570 (electronic)
Biosystems & Biorobotics
ISBN 978-3-319-97766-9 ISBN 978-3-319-97767-6 (eBook)
https://doi.org/10.1007/978-3-319-97767-6

Library of Congress Control Number: 2018950824

This Springer imprint is published by the registered company Springer Nature Switzerland AG
The registered company address is: Gewerbestrasse 11, 6330 Cham, Switzerland

To my father,
absorbed on the typewriter,
steadfast gaze and smiling,
traveler in the name of
Agricultural Law

Preface

"The mark of science is in its teaching". I learned this very early, from my father, a rigorous and watchful scholar of Agricultural Law, who wanted to certify the autonomy of his discipline as progress and distinction with respect to the others that existed at the time. It was from his interpretation of Aristotle that he drew the motivation that summarized his effort to write a book for students. Without a manual, a text with a methodological systematization of knowledge, it is not possible to train students and build a discipline within which a scientific community can be developed, and which can then be reflected in the programs of university courses and research areas.

Destiny has decreed, after almost 20 years since my father's death, that I have found myself, by experience and by pure chance in life, working in a relatively new and original sector. This is biomedical engineering, in particular biorobotics, which requires the same systematic activity in order to become well established.

But today, we are in the years when the scientific disciplines, and consequently their teaching, are undergoing an impetuous evolution, and the age-old and slow effort to divide them into separate sectors is being constantly tested by the emergence of new methods and forms of knowledge. These are transforming the very method of teaching through manuals and textbooks, and they endorse a drive toward interdisciplinarity, or rather, *antidisciplinarity*.

The tendency to cross and break down barriers stems from the request to address the contingent problems of humanity manifested by public opinion and politics in relation to the scientific community. This is because an increasing number of skills are required that cross the various scientific areas in order to resolve the problems of a general nature, and the solutions are never confined to a single discipline. In order to face the problems of society, it is necessary to federate between scientists of various sectors and to collaborate in programs aimed at involving more areas. On the contrary, if we consider the needs of teaching, the stimulus is in the opposite direction. To train, we need to compare and deepen, systematize, and build a discipline with its methodologies, so as to be able to prepare the scholars and professionals of tomorrow. Who teaches knows well that the training must be thorough and rigorous, and be based on the "fatigue of the concept" in order to allow a true transmission of knowledge to those who one day will not only make it theirs, but who will also modify it with their own original contribution.

Within this context, this book is the first step toward an attempt to present an area, that of robotics, which is studied in the most advanced engineering courses, and attracts a great deal of attention as a scientific and technological field. Not only is this due to its various applications and current implications, but also for its potential to revolutionize our life.

It is therefore a journey through robotics on its way toward its evolution from industrial to social applications. This book originated from a series of lessons that I held at the School of Politics in Rome, during 2016. The lessons represented a landing point, at the end of a transitional period that I experienced with my dual spirit of scientist and politician, following my return to Parliament after the experience of government in 2014. In fact, in a very difficult phase of life, I reinvented my research activity and the parallel political initiative, attempting to integrate them in an original proposal, starting again with study, and wishing to constructively close my chapter of Ministerial experience.

There is nothing better than a break from work, a period of sabbatical like the extraordinary one that I experienced in the government, to return to study and reflection with a new stimulus and with the spirit of observation that only a detached perspective can bring. Studying and preparing to reenter the world of robotics with greater awareness and maturity, I came across a veritable explosion of literature on the subject. This was not only purely scientific but of various kinds, including science fiction, journalism, education, and economics, which all showed a new vitality of the industry, a new life for robotics, with a huge impact on society. I immediately realized that in the course of just two years the world had changed profoundly, especially from a scientific and technological point of view. We were entering a phase of profound transformation in the production of goods and services, with a capacity to change society, work, and our way of life. From 2014 to 2016, I then tried to observe this transformation, to understand the real effects, more than the theoretical or potential ones, of what is commonly called the "fourth industrial revolution". In this period of research and reflection, I evidently observed the industrial world from the perspective of robotics and bioengineering, my scientific sphere, in order to analyze phenomena, scientific and technological developments, and the impact on our world.

A little later, I was invited to some universities and research centers in Italy and abroad, where my contacts had taken me. It was very important because comparison and relationships with colleagues are an essential part of my research career and have always enabled me to give structure to my primitive conjectures by verifying ideas together with others, just as knowledgeable as myself, and more so, in various disciplines influenced by the fourth industrial revolution. I went to public and private research centers such as Imperial College London, Ericsson in Stockholm, Waseda University in Tokyo, the Italian Institute of Technology in Genoa, and the Istituto Superiore di Sanità in Rome, but in parallel with these prestigious centers, I also came into contact with many schools and universities. The students, even the very young ones, welcomed my somewhat out of the ordinary lessons with great willingness, interest, and critical spirit.

Inevitably, the cycle of lessons and its mutation with time have been affected by my journey through different cultures and geographical areas and have evolved along the road to the pathway that inspired this book. Along this road I found myself faced with students of very heterogeneous and varied origins, offering them an incentive to understand the transformation underway, and proposing ideas of appropriate public policies in the field of economic development, training, and research policy.

My "migration" to conferences and meetings from South Africa to Japan, from Serbia to Iceland, allowed me to compare myself with others, to integrate, to add a good deal of substance to the quality of my lessons, and to prompt me to have different and deeper perspectives for analyzing the phenomenon of the fourth industrial revolution and its contextualization in robotics and biorobotics. In parallel with the lessons, I was also involved in supporting a start-up of the Scuola Superiore Sant'Anna in Pisa. This meant I had to discover the pitfalls of enhancing research, and the transformation of science into products and markets, in such a competitive and difficult context like the one created by the emerging fourth industrial revolution. I think that this experience, such as the one in European research policies, as President of the Panel appointed by the European Commission to evaluate the impact of research programs on technologies of the future (FET Flagship of Horizon 2020), has influenced my way of presenting and describing problems, which now also reflects the industrial, political, and social experience I have gained.

The book should therefore be seen as an intermediate step in my "practical" course of study and has a strong motivation linked to the attempt to verify the possibility of the "mark of science" and therefore of its teaching.

This is why I decided to transform these lessons into a book which, far from being merely a manual, represents a notebook, written and meditated, to arouse debate, a discussion on the nature of this revolution and on the role of robotics.

The transcription of the lessons into a systematic text was totally new to me and represented a certain effort in setting to paper something that is very much in motion and requires reflection and sedimentation in order to be transformed into a story with a narrative continuity. Italian author Claudio Magris describes, in the preface of *L'Infinito viaggiare* (Endless Journey), exactly how I felt: "So it happens with writing; something which, while traveling and living, seemed so fundamental has vanished—on paper it is no longer there, while what imperiously takes shape and imposes itself as essential is something that in life—in the journey of life—we had hardly noticed". Likewise, and much more modestly, this book is different from the oral lessons that I held at the School of Politics. With their transformation into written form came ideas and reflections which have shaped the substance and form of my journey in robotics, and I hope to inspire curiosity and desire in the reader to deepen their knowledge further. The book is addressed to all readers, who, starting from their training, not necessarily scientific, want to understand something more about robotics and the industrial revolution. It is not at all necessary to be a scientist in order to grasp the messages that the book wishes to launch.

Finally, I would like it to be used to draw attention to the changing world, what is happening in Europe and in the rest of the world. The only way to face the fourth industrial revolution without being overwhelmed is to study it, understand it, and make ourselves protagonists and actors, and not consumers. As former Minister of Education, I aspire to an Italy as cradle of cultural, scientific, and technological ferment, certainly not limited to the market merely for the profit of others.

The only way to gain this destiny is to invest in education and training. We must change the school system and its syllabuses and continue throughout our life to educate, update, and train our citizens to prepare themselves to be aware and protagonists of their lives, and therefore in full control of themselves and their own future. The take-home message is therefore beautifully simple and crystal clear. It is necessary to study and prepare ourselves so as not to be caught unawares in our backwardness, above all to be always motivated in our desire to become creative protagonists.

Pisa, Italy   Maria Chiara Carrozza

Contents

About the Author

Maria Chiara Carrozza is Professor of Industrial Bioengineering at The Biorobotics Institute of Scuola Superiore Sant'Anna, Pisa, Italy, where she served as Rector from 2007 to 2013. Since January 2018, she is Scientific Director of Fondazione Don Carlo Gnocchi an Italian Research Hospital dedicated to Rehabilitation and Personal Assistance. She currently serves as President of the National Group of Bioengineering. She has teaching and research experience in several Institutions in Italy, Europe, USA, Japan, Korea, and China. She is also a member of the Robotics and Automation Society (IEEE-RAS) and of the Engineering in Medicine and Biology (IEEE-EMBC). She was principal investigator and coordinator of several international research projects, funded by the European Union. From 2016 to 2017, she was Chair of the Interim Evaluation of the Future Emerging Technology Flagship Program and Member of the Steering Committee of the Quantum Technology Flagship for the European Commission. She also holds several patents and is involved in projects for supporting technology-based innovation. She is Founding Partner of IUVO srl, a start-up active in Wearable Robotics. Since 2015, she is Independent Director of Piaggio SpA. She has been a Member of the Chamber of Deputies (*Camera dei Deputati*) of the Italian Parliament from 2013 to 2017 serving in the Committee of Foreign and European Affairs. She was Minister of Education, University and Research from April 28, 2013 to February 22, 2014.

1

Are We Going Through a Real Revolution?

It is not easy to understand if what we are experiencing is truly a new industrial revolution or a tail or a transformation of the previous one, the third revolution, which dates back to the late seventies and was originated by industrial automation, microelectronics and mechatronics.

In an attempt to understand what is happening today in terms of the transformation of society and of the industrial revolution, it is necessary first of all to go back to the industrial revolutions,[1] to their powerful impact on society, to the discontinuities that scientific discoveries and their subsequent transformation into technology and industry have produced over the years on society. This is not only on the manufacture of goods and services, but above all on the organization of work and on the quality of life.

Without encroaching too much on the work of historians, it is necessary to understand the history of the industrial revolutions and the mechanisms that have enabled their evolution, as well as the economic policies that have favored them. Above all, it is important to know the relationship between science and politics and between innovation and social changes in order to foresee what may happen also in the current fourth industrial revolution.

Faced with the current machine revolution, which will lead to a progressive transformation of our world, each of us may very well have two reactions. The first is refusal, which can be active—by participating in movements that oppose or try to delay this evolution (the so-called *neo luddism*[2])—or passive,

---

[1]Robert Allen (2017), the industrial revolution. A very short introduction, Oxford University Press.
[2]In English sometimes it is called 'reformed luddism', to know more see https://www.ksl.com/?sid=23241639.

© Springer Nature Switzerland AG 2019
M. C. Carrozza, *The Robot and Us*, Biosystems & Biorobotics 20,
https://doi.org/10.1007/978-3-319-97767-6_1

maintaining a detachment from what is happening and operating a refuge or return to activities that presumably will not be greatly affected by events. The second reaction, very desirable and positive, is instead that of wanting to be an active part of the ongoing process and for this reason equipping ourselves intellectually and cognitively to be creative protagonists.

Consequently, for those interested in the future of Italy, it is essential to try to work on the development of policies to prepare the country, its workforce and its young people to face this challenge. Being ready is everything: if we want to remain a manufacturing country, we must be an active part of this revolution, create businesses and train citizens capable of living to the full and being participative creators.

Where is the paradigm shift necessary, according to the history of economics, to trace the sign of a real industrial revolution?

The economists identify enabling and transversal technologies as the key elements that determine discontinuity in the working method, the change of production and how these reflect on society and civil life.

Two clear examples of enabling technologies of the first industrial revolution, at the end of the eighteenth century, are the steam engine and the automatic loom. These inventions produced fundamental changes in the organization and management of manufacture, respectively replacing the manual energy source with steam power and making the weaving process faster for the benefit of productivity. These were two historic transitions that profoundly affected the conversion of energy and the automation of the production process.

The United Kingdom was the country to benefit from owning key technologies and becoming the first industrial power in history.

The second industrial revolution was driven by the scientific discoveries of the last part of the 1800s, of which some examples are electricity, steel and chemical fertilizers. Germany also became an industrial power, possessing such enabling technologies thanks to investments in research and innovation, as did the United States, where at the beginning of the 20th century the Ford motor vehicle assembly line was invented, schematized and portrayed in *Modern Times* by Charlie Chaplin.

Coming to more recent times, the third industrial revolution arose around 1980, and is linked to microelectronics and the industrial automation closely associated with it. The propulsive drive of the third industrial revolution is well described, from a technological point of view, by the so-called Moore's

Law.[3] In short, this provides for a doubling of the number of transistors in a single chip every two years, and the progressive increase in the speed of execution of commands, by virtue of constant technological investments. The consequence of this has been to produce ruthless competition among truly giant monopolists of enabling technologies. Technological development has therefore followed Moore's Law, which periodically has foreseen investment in equipment and machinery in order to keep pace with the increase in transistor density in a chip, with the side effect of involving a Darwinian selection of industrial players. Ultimately, the third industrial revolution has given us a world of large multinationals holding intellectual property and know-how to produce consumer electronics. They themselves are the only engines of technological progress, and manage or incorporate innovation by 'acquiring' people and startups. As an example it is worth pointing out that the Italian electronics industry has been one of the victims of this evolutionary scenario and has remained excluded from the microelectronics and ICT race due to its inability to follow this veritable whirlwind of technological innovation.

One of the determining factors in the transitional phase that we are experiencing is precisely linked to the context in which the third revolution seems to have reached its maturity, associated with the potential end of Moore's Law.[4] In effect, it is possible that the propulsive drive to microelectronics and the increase in processing speed is coming to an end. As a consequence, capital will have to be engaged in the transformation of scientific discoveries into innovative technology so as to fuel a subsequent revolution, and trigger a new cycle of industrial expansion.

As was the case in previous industrial cycles, today some countries and their large industries are observing the evolution we are going through, trying to impose their leadership, and thus becoming masters of the key competences and technologies that we presume will be decisive for the future. The transformation of science into technology—and its use in processes and products—are the crucial phases of the industrial period. Whoever dominates and possesses the technologies wins.

Within this context, prompted by the German Academy of Sciences, Engineering and Industries, in 2011 the term Industry 4.0 was coined.[5] This was to produce a label aimed at indicating, rather than the observation of

---

[3]G. E. Moore, *Electronics 38*, 114–117 (1965).

[4]M. M. Waldrop, *The Chips are Down for Moore's Law*, Nature 530, 144–147 (11 February 2016).

[5]*A fourth industrial revolution is about to begin (in Germany)*, «Wired», March 21, 2015, http://www.wired.co.uk/article/factory-of-the-future.

a historical phenomenon, a national program that the German government has decided to support to address the epochal transition we are experiencing in the field of technologies for the production of goods and services. Similarly, also the Italian government, urged by Parliament, has activated, albeit with some delay, an Industry 4.0,[6] which aims to support the renewal and transformation of Italian industry in view of the fourth Industrial Revolution.

In the Industry 4.0 program, among the various actions envisaged, the enabling technologies that are expected to be decisive are clearly listed. Some examples are robotics, three-dimensional additive mechanical technologies, cloud and telecommunications. It is therefore assumed that all together, and integrated, these technologies produce a discontinuity that will revolutionize our industry.

So, even if it is not possible to anticipate historical analysis, we can be very sure that European governments are convinced that a research program and the right investments are needed to face the transformation of our society following the fourth industrial revolution. The governments then gave a name to this phase, and they mobilized to produce a program known as Industry 4.0. Research centers and enterprises will have to develop knowledge, intellectual property and key enabling technologies that are crucial for the future.

But what really distinguishes the transition from 3.0 to 4.0? Is it a real leap from 3 to 4, that is, a clear technological discontinuity with facts and figures, or is it merely a communication ploy to give a name to a research and development program? Is it a real change or simply something strongly desired?

In short, why do we say Industry 4.0 and not 3.1, 3.2? Are we just carrying out a software update or is there something deeper?

Ultimately, it is all about understanding if the world of industry is changing so profoundly that it can be contextualized within a new industrial revolution and whether or not we can really expect something new and earthshaking. The key to understanding this transformation is robotics because it is one of the enabling technologies around which we can envisage the revolution taking place. We will try to demonstrate why robotics can be the symbol of this transformation, so first and foremost we need to understand what a robot is and what it is used for.

Introducing robotics is particularly complex. Although we might very well have a clear image of robots, we are facing a phenomenon in complete

---

[6]Industry 4.0 Plan, Ministry of Economic Development http://www.mise.gov.it/index.php/en/202-news-english/2036690-national-industry-4-0-plan.

evolution that regards a non-discipline, which in itself cannot be defined in a univocal and scientifically repeatable way.

First of all, it is necessary to understand what robotics is, how it has spread and how it has already profoundly influenced industrial production, radically changing assembly lines and making possible, through automation, an acceleration of production times, high precision and high quality such as to make our world the way it is: uniform and globalized in terms of products and services.

Starting from these considerations, we can say with certainty that robotics already made its contribution to a significant change around the eighties, profoundly overturning the mechanized "assembly lines" of the early 20th century Fordist revolution and production lines, driving the productivity of manufacturing plants to the level of the present day. This was the so-called third industrial revolution, which occurred around the 1980s, originating from the combination of different but intensely intertwined elements—microelectronics, mechatronics, robotics and industrial automation.

To describe its creative potential, we must familiarize ourselves with some concepts. The *robot* is basically an intelligent physical system, equipped with actuators and sensors. It has artificial intelligence, linked to software and learning systems, which process information and allow some forms of reasoning, but above all provide the capacity for autonomous decisions without the direct control of a human operator. It uses the virtual *cloud*, located on remote servers, which contains the library of previous information and also that acquired by the sensors of the robot, making the perceptual experience both shared and shareable between a number of robots. Thanks to the enhancement of telecommunication technologies, the cloud becomes accessible in real time, so enabling the sharing of decisive information for movement in unstructured and unknown environments. Here the robot can navigate thanks to the data library that is accessible without significant time delays.

The fourth industrial revolution can therefore act in two directions, determining:

(a) an *impact on the manufacturing world*, because the production of goods and services by means of robots, artificial intelligence, telecommunication technologies and the cloud can be completely reformed and modified;
(b) the *transformation of society* because the entry of the 4.0 robot will take place very much in our midst.

The first trend will have disruptive effects. The cloud empowers robotics, because it creates collective intelligence and shares perceptual experiences among different robotic agents distributed in the environment, for example allowing the robot to download perceptual experiences of other robots and information useful to navigate in unstructured environments and manipulate. In practice, the robot acquires physical skills, the ability to process information, make decisions and conserve memory and knowledge in the cloud.

In essence, the 4.0 robots can move better, be more versatile and autonomous in carrying, manipulating or moving objects. They can communicate efficiently with each other, transmit useful information and share cognitive experiences in the cloud. However, it is the entry of robots into society that influences the imagination and represents the real challenge. In fact, the approach of the robot towards people can only take place once it is able to understand and interpret their desires, through an appropriate human-machine interface, and above all when it is safe in its physical interaction.

As this step is critical, we need to focus on integrating physical abilities, cognitive skills and the collective intelligence available through the cloud. In order to clarify how these qualities can make the robot better than it is now and make it more "efficient" in performing tasks in a functional and satisfying and at the same time "safe" way. The objective of research in view of the present industrial competition means precisely overcoming the current limits, to make the robot truly sentient with respect to people, executing tasks without temporal latency and secure in its physical interaction with them. This is where the heart of our studies lies and it is fascinating to discover how simple this concept is to describe and yet at the same time so difficult to achieve.

In fact, we have already highlighted that the first fundamental contribution of robotics to an industrial revolution has already taken place, concerning the third industrial revolution. It took place in the eighties of the twentieth century and involved manufacturing, and above all industrial automation. Just think of the robots on the assembly line for the production of vehicles, often portrayed in advertisements and videos. The power, speed of working and quality are impressive and have enabled the production of a new vehicle in a rapid, precise and repeatable way in a few minutes, reducing the margin of error and the relative production defects. Industrial automation uses robots for handling and manufacturing parts in a production line, and artificial intelligence and information technology to support and manage the robots. The fundamental issue is that at this moment throughout the world little can be mass-produced without the aid of industrial automation. Everything else is handicraft.

So what is automation? The word itself already makes the concept explicit. It is enabling some processes to be automatic and independent of human action, but at the same time making them controllable, repeatable and efficient.

Industrial robotics has therefore been the technological basis that has made it possible to revolutionize manufacturing, has replaced human hands and arms in repetitive and high-frequency or tiring tasks, and has come a long way to supporting and sustaining production. While it has supported the work of workers in production, eliminating their tiresome, weary or repetitive tasks, on the other hand it has certainly taken away low-profile jobs and has led to a transformation of the manufacturing environment.

The progress of an economy or nation today is also measured by considering how many industrial automation robots are installed compared to number of workers. This ratio is an indicator of the level of competitiveness and technological maturity.[7] According to the International Federation of Robotics, in 2015 in Italy there was a density of 160 robots per 10,000 employees, compared with the density of 301 for Germany or 501 of South Korea, and thanks above all to investments in China, there is expected to be a surge in the number of robots installed worldwide, which is expected to reach 2.6 million by 2019. There is therefore a first problem already present in all industrial revolutions, but which in the fourth will be particularly disruptive, that is the relationship between technological evolution, automation and jobs. We can imagine that in the fourth industrial revolution a lot more could very well happen, that the robots will come out of the factories and come among us, entering the service industry and our homes.

Let us now face the paradox constituted by the fact that everyone knows or thinks they know what a robot is. In fact, robots are described in such a widespread manner and displayed in films, literature and comics that we risk not being able to transmit in appropriate depth the scientific reality and complexity. It is also difficult, after having seen the imaginary robots in literature, for our real robots to meet our expectations because in their construction they are always primitive with respect to the performances that are attributed to them in literature and cinema.

The clear fact that literature and art have "preceded" science is the most interesting aspect of this work. It is increasingly common for me to give lessons on robotics to a mixed public, and it has never been difficult to get across what a robot is, and what it can contain, because it is like describing a biological system and going to the roots of intelligence itself. Even we scholar-designers of robots, scientists and technologists, who in every moment

---

[7]*World Robotics Report*, International Federation of Robotics (IFR), http://www.ifr.org/news/ifr-press-release/world-robotics-report-2016-832/.

of our life waver between the fascination of science, the passion for technology, and the desire to become producers and entrepreneurs, sometimes treated as "enemies", sometimes as gurus or great thinkers, have difficulty in finding univocal and simple words to define a robot.

So anyone can understand and intuitively gain an idea of what a robot is, but this widespread perception makes it difficult to define scientifically what it is, without "crossing" the border and confusing a robot with an intelligent machine, a mechatronic system, a system of artificial intelligence. In short, this is the great fascination of robotics.

One example of a robot, already present in our homes, is Roomba,[8] an automatic vacuum cleaner that has been sold in its thousands. It automates the robot's navigation in the house and carries out cleaning without the need of a human operator. How many of those who have bought the Roomba are aware of the logical leap that they have made, bringing home a robot that makes cleaning independent?

Speaking of robots and their influence in society is therefore not particularly new and is also strongly reflected in all forms of artistic and literary expression, to the point that one can speak of "anticipatory" literature of the social implications of the industrial revolution. If we wish to understand the sociological implications, we can start from books, comics and films, which have already said a lot, effectively almost everything, regarding the charm, fear, risks and the great attraction of the construction of the "artificial otherness" produced by the development of robotics. The father of literary robotics, Isaac Asimov,[9] predicted many of the problems and social implications we are facing today with the entry of the robot into society. Cinema revealed in advance, with *Blade Runner*,[10] the aesthetics and identity of the replicating robot and the difficulty of integrating moral laws[11] into robots. In the film, once adequate moral laws are introduced that enable robots to operate successfully in society and interact with people, they also develop a kind of melancholy and sense of inferiority for not having memories, not to be "complete" men and women. They also feel treated unjustly. It is almost as if the awareness of existing and the ability to make decisions, encroaching on a kind of consciousness, then lead the robots to be humanized but at the same

---

[8]Roomba is produced by IRobot, an American company founded by Rodney Brooks, a veritable genius of robotics and technology transfer (http://www.irobot.com).

[9]I. Asimov, *I Robot, 1950.*

[10]*Blade Runner*, 1982, film directed by Ridley Scott; adapted from P.K. Dick's novel, *Do Androids Dream of Electric Sheep?*

[11]The three laws of Robotics are defined in the 1950 collection *I, Robot* by Isaac Asimov.

time to feel incomplete and in competition with humans, who are perfect as owners of memories, of childhood and family ties. These parts of the film, in particular, introduce the question of robotics and moral laws, investigated in "technoethics",[12] an interdisciplinary area that studies the relationship between ethics and technology. There are countless other literary and cinematographic sources regarding these issues, especially on the interrelationship of humans with technology and automation. It is suffice to remember *Mr. Hulot* in *Mon Oncle*,[13] a recently restored cult movie of the French cinema or Alberto Sordi in the Italian film *Io e Caterina*.[14]

If those described above are problems, or frustrations, generally contextualized in the relationship between man and robotics, widely depicted in advance in the 1950s, in reality the fourth industrial revolution can represent an quantum leap in quality compared to the past. Indeed, it will favor the entry of the robot into society and the creation of the relationship described in *Io e Caterina*, and thus make it essential to introduce moral laws for robots, to endow them with the ability to work in the midst of humans. Literature and cinema have foreseen these issues and have given us the opportunity to reflect on the processes in order to make ourselves aware of the difficulties we are facing.

However, what do we not know? The novelty—probably the most critical and difficult part of the path that awaits us—concerns which technological and scientific advances have produced a potential discontinuity that can enable robots to improve their performance and if and how this improvement will bring the robots to work in our midst, making them compatible with our presence in a safe manner.

In short, it is an industrial revolution that presents us with a world in which intelligent machines will enter society. The machines will help or even replace workers in producing important services, such as those provided by care givers to persons. Therefore, it potentially could be a "negative" revolution that excludes and does not involve, which renders whole classes of workers and entire classes of people useless or obsolete. If we look at the individual worker, it is difficult not to think that a part of their work, or even their entire job, will soon be performed by an intelligent machine or robot,

---

[12]J. M. Galvàn, *On Technoethics*, http://bib26.pusc.it/teo/p_galvan/technotehics%20ieee%20tradotto.pdf.

[13]In the 1958 film *Mon Oncle*, the protagonist Mr. Hulot, a character played by Jaques Tati, "discovers" an automatic kitchen but is unable to either understand or use it.

[14]*Io e Caterina* is a 1980 film directed by Alberto Sordi, in which the protagonist has a humanoid robot assistant at home.

which will take charge of a series of activities independently thanks to personalized cognitive and physical skills.[15]

The change that will produce the fourth industrial revolution, however, may represent a historical discontinuity with a potential impact on our world. This will be very marked because the robots will come out of the factories and will enter our society to work in our midst. This is why the symbol of the fourth industrial revolution is undoubtedly the driverless car,[16] because in it there is a concentration, as a metaphor, of all the implications and meanings of the revolution itself. The entry of the robots among us, on the streets, the replacement of the driver with sensors, actuators and a control system that automatizes driving and therefore decides the behavior of the car, all represent a change in the business model of urban or extra-urban private mobility. The economic system will therefore change profoundly. Both the car sales market, and the driving and mobility service for private use will be revolutionized by innovations. The self-driving car thus becomes emblematically the symbol of this revolution already underway, which makes it necessary for car companies to invest in research and development, with strategic and commercial agreements between consolidated companies and startups. This will also make it possible for the large digital companies to make investments in and very probably modify what is an imposing industrial sector ("the industry of industries", according to Peter Drucker[17]) involving hundreds of thousands of production workers, and represents an economic backbone in many states.

It is also important to underline that the driverless car—and the research associated with it—means for the first time extensive legislative intervention. This involves the normative and regulatory bases that allow the use on city roads and motorways of systems and algorithms that autonomously take appropriate decisions to avoid damage to property and persons. It is a significant challenge for our laws[18] and above all for our ethics. In fact, for the first time moral laws and the ability to discriminate between various situations, must be placed in an intelligent machine to enable it to operate in the midst of humans and their possessions. It is not just a question of insurance or liability, but much more. We have spent years in which we thought

---

[15]A. McAfee, E. Brynjolfsson, *The Second Machine Age*, W. W. Norton & C. Ed., New York, 2014.

[16]The historic forerunner project on autonomous driving is that of Google https://www.google.com/selfdrivingcar/.

[17]P. Drucker, *Concept of the Corporation*, John Day, New York, 1946.

[18]Important work was carried out by an ad hoc European Parliamentary commission which elaborated a document with recommendations to the European Commission on the application of civil law to robotics. http://www.europarl.europa.eu/sides/getDoc.do?pubRef=-//EP//NONSGML%2BCOMPARL%2BPE-582.443%2B01%2BDOC%2BPDF%2BV0//EN.

machines should be "stupid" and only obey orders. Now we find it necessary instead to transfer our morality to the machines in order to enable them to make decisions in our place, how to stop the vehicle at the right time, avoid running over a pedestrian, preserve the well-being of those on board or safeguard those on the street. To understand this phenomenon, it is suffice to see how supporters of the utility of driverless cars highlight the aspect of safety with respect to road accidents, arguing that with the automation of private vehicle driving there will be fewer accidents and greater overall safety.

We have therefore seen how the scenario of the fourth industrial revolution is still to be written. The complexity of the current socio-economic framework is different from that of the past. Robotics and artificial intelligence will be candidates as technologies enabling the future only if they are able to meet the challenges related to the specific social, legal and moral contexts, but also the ever more pressing requests for economic, energy and environmental sustainability. The pathway towards obtaining sustainable robots poses technological and scientific problems that make the picture far more complex than that of previous industrial revolutions, demanding appropriate investment to support the enormous challenge we are facing.

2

On the Way to Robotics

I do not intend to describe the history of robotics[1] and its literary origin, dating back to the beginning of the twentieth century,[2] but I would like to trace the technological evolution of its applications, imagining a journey through time and space.

My analysis strongly reflects my training as a bioengineer researcher[3] who has always delved into the interaction between robotics and biomedicine. This is why I think it is useful to take a step back and describe my studies.

I do not have a formal robotics background, also because at the end of the 80s it was not yet a part of my future plans. As a scholar of physics, I learned and studied in research laboratories, in a "middle world", on the border between microengineering, robotics and biomedical engineering and I think it is for these reasons that I have taken on board a research methodology oriented towards antidisciplinarity and guided by curiosity.

The passion for robotics was therefore born later than my academic training. However, I gradually specialized in research fields to this scientific

---

[1]B. Siciliano, *Handbook of Robotics*, O. Khatib Editors, Springer-Verlag, 2016.

[2]The word "robot" originates from Karel Čapek's 1921 play entitled, *RUR* (Rossum's Universal Robots), from the Czech word "robota" which literally means "hard labor".

[3]For a definition of bioengineering see Encyclopaedia Britannica https://www.britannica.com/technology/bioengineering and the definition of biomedical engineering research at the web site of the IEEE society of Engineering in Biology and Medicine (EMBC) https://www.embs.org/about-biomedical-engineering/what-we-do/.

© Springer Nature Switzerland AG 2019
M. C. Carrozza, *The Robot and Us*, Biosystems & Biorobotics 20,
https://doi.org/10.1007/978-3-319-97767-6_2

area, up to the most recent part of my scientific career, where I became the founder and animator of a sector that can be defined as neurorobotics. I graduated in elementary particle physics and I did an internship at CERN[4] for six months in order to participate in the experiment on which I then wrote my thesis. It was an experiment of high energy physics, aimed at studying the violation of CP[5] in the decays of K mesons.[6] My contribution was a small cog in a complex experiment that involved dozens of people. Being in Geneva at CERN, in an international environment, with colleagues from all over the world, all brilliant, zealous and serious, who spoke the universal language of science in English was what helped me to understand that my world was that of research. In addition, the internship in Geneva also helped me to develop humility and respect for those who knew more than I did, and to familiarize myself with the discipline of international organizations, which characterizes the great scientific collaboration projects and made me able to listen and absorb everything I saw happening and being discussed around me. I felt a strong admiration, and at the same time the ambition to be like the scientists greater than myself. I wanted to become a respected researcher and study, learn and experience the life of experiments and analysis of data.

Upon returning from my CERN experience, I graduated, started a family, and I presented as many applications as possible for access to PhD Programs.

Between competitions and interviews I learned that in scientific or professional life we are not the ones to choose, but there are others who choose us and that we must be flexible and adaptable in order to be chosen. I was admitted into the PhD Program in Engineering of the Sant'Anna School of Advanced Studies, which opened a new research path for me—that of microengineering for biomedical applications. It was a pioneering area, not yet defined academically but promising, because it was the protagonist of the third industrial revolution as it was derived from microelectronics with an extension to mechanics. It was Prof. Paolo Dario,[7] my first real mentor, who suggested that I should join the Sant'Anna School's microengineering laboratory. The research goal was to open new frontiers in silicon microfabrication technologies, which were at that

---

[4]CERN is the European Organisation for Research into Nuclear Physics that has its headquarters in Geneva, in Switzerland. For more details, see https://home.cern/about.

[5]Violation of the symmetry of charge and parity: one of the aims of elementary particle physics is the study of CP-symmetry and its violation. For a detailed volume on the topic, see D. H. Perkins, *Introduction to High Energy Physics*, Cambridge University Press, 2000.

[6]K mesons are elementary particles, D. H. Perkins, Op.cit.

[7]Prof. Paolo Dario, Director of the Institute of Biorobotics of the *Scuola Superiore Sant'Anna*, is one of the most cited Italian scientists. His Google Scholar page gives his main publications https://scholar.google.it/citations?user=dW7Fi88AAAAJ&hl=it.

moment undergoing the transition from purely microelectronic to microme-chanical characteristics and my physics skills would integrate well with those of mechanical and electronic engineers. The mechanics, the dynamics and the fluid dynamics of the micro-world are completely different from those of the macro-world.[8] A physicist like me could "be of use" to fine-tune the experiments, for the measurements and for the integration of the traditional knowledge of mechanical engineering. Paolo Dario had great intuition, which was to imply a profound change not only for those who worked in our laboratories, but also for the history of the Sant'Anna School of Advanced Studies because I sincerely do not think that our university would be what it is now without robotics, mechatronics and micro-engineering.

In a research environment organized in small groups it was possible to be protagonists of your work, focusing on the production of microsystems and devices to be applied in medicine, with a direct impact on the quality of therapy and patient care and with a possibility of very rapid industrial exploitation. It was precisely this combination of quality international research, development, social innovation, progress in medicine and industrial exploitation that fascinated me and made me enthusiastic about working in microengineering. My PhD program was therefore in mechatronics, and in particular micro mechatronics for biomedical applications, such as the delivery of drugs in controlled doses with miniature, and possibly portable, devices.

What is mechatronics? Defining it helps us to reach robotics from another angle and enables us to better explain the third industrial revolution, fol-lowing its geography, because mechatronics was above all Japanese as well as German and European. Mechatronics[9] is a very powerful technology and originates from a term invented in Japan that indicates the fusion between mechanics, electronics and information technology and emphasizes the integration into a miniature device of a mechanical system containing actu-ators, motors, sensors and the relative controller, in order to achieve complex and intelligent operational characteristics.

The prototype of a mechatronic system, which is also illustrative of Japanese brilliance and the ability to transform a concept into a philosophy of design and industrial production, are the automatic cameras that were

---

[8]R. Feynman, *There is Plenty of Room at The Bottom*, Dec. 1959, is a talk by Nobel prize winner Richard Feynman, who explains for the first time, in a marvelous way, all the physical characteristics of the micro-world that make it so fascinating and interesting also for engineering. For a transcript, see: https://scholar.google.it/citations?user=dW7Fi88AAAAJ&hl=it.

[9]For a definition of mechatronics, see the Journal of Mechatronics, https://www.journals.elsevier.com/mechatronics/.

developed starting from the mid-seventies by Olympus.[10] The innovation, born from the idea of making reflex cameras more compact and accessible, went on to bring about a true democratization of photography through automatic regulation. Thanks to the integration of sensors, motors and the control system, they could automatically adjust parameters such as exposure time and focus to environmental conditions. In this way, the ability to take quality photographs, which was previously only possible with traditional cameras and with professional photographers behind the machines, was available to everyone. There were small inexpensive and efficient automatic machines that enabled non-experts to take high quality photographs. The power of mechatronics lies therefore in the integration of mechanics, electronics and control, in the miniaturization of the system and mechanisms and finally in making the adjustment of mechanical and electronic parameters automatic according to external conditions. Mechatronics has been and continues to be extremely powerful also in Germany and Italy, but we can say that it has Japanese roots, because it enabled Japanese industry to compete in integrating microelectronics with micromechanics, exploiting the skills in precision mechanics and associating new features to products thanks to the use of microelectronic components. There are many other components that have a different geographical characterization, such as Silicon Valley, but which are just as symbolic of an industrial revolution that from a purely micro-electronic angle gradually became micromechanical and micromechatronic. The best-selling component that has profoundly affected the automotive industry is the accelerometer contained in the airbag of cars. The accelerometer[11] is an inertial sensor that measures acceleration and has the purpose of inflating the air bag, a device that provides security and deadens the impact for passengers in the case of an automobile accident. It is an integrated chip that contains on board not only the electronic part and conditioning and control of the signal, but also the mechanical component made up of the oscillating mass that measures the acceleration due to an accidental impact. Now in a single mass-produced chip, silicon accelerometers are created that are set up in all the vehicles produced. Here is the power of the third industrial revolution. Microelectronics—and subsequently micromechatronics—enable the mass-production at a limited cost of integrated microsystems that carry out measurements or realize complex

---

[10]To see the various Olympus models, in particular those produced in the period 1970–80, see https://www.olympus-global.com/brand/museum/camera/.

[11]M. Aikele et al. *Accelerometers For Airbag* in *Applications Advanced Microsystems for Automotive Applications 99* (2013): 251.

functions containing mechanical and electronic parts merged with each other with a new design approach based on the enabling technologies of silicon micromachining.[12]

If we talk about medical applications—which I studied in my PhD thesis and which I dealt with in the following years—we must refer to the *biomechatronics* sector. Two examples illustrate the innovative impact biomechatronic approach: the innovation of system for laparoscopy[13] in the broader area of minimally invasive surgery, and miniature wearable devices for drug delivery via implantable micropumps, for insulin delivery therapies. Many other applications of medicine require precision mechanics combined with sensors and dexterity of movement in small environments, such as the inside of the human body. In my PhD thesis I studied stereolithography for microfluidic applications, that is, to create micro-pumps for fluid dosing, using 3D manufacturing microtechnologies. This was exploring what would later become the promising field of three-dimensional molding technologies, which began to develop during the 1990s.

In my postdoc years, after having had some contracts with the European Space Agency where I used my knowledge of microfluidics in the microsatellite field, I started, at the suggestion of my supervisor Paolo Dario, to work in the endoscopy sector, in particular regarding the colon. I have a strong bond with the colonoscopy sector, which enabled me to work together with colleagues and experts and help develop the so-called "robotic colonoscope".[14] This is an instrument which, through propulsion inspired by inchworm locomation, is used to move a camera and microinstruments through the colon for operational colonoscopy. This is with the intention of replacing, wherever possible, the current manual colonoscopy, which is very painful because it is "pushed" into the colon by blowing air and causes painful contractions. Precisely the "robotic inchworm" highlights the transition from mechatronics to robotics, because the onboard sensor and intelligence are combined with features of navigation and manipulation. Therefore, as we reach the ability to move independently, we can say that it is no longer a mechatronic system but a robot. As often happens in these areas, it should be noted that the boundary between the two worlds, mechatronics and robotics,

---

[12]Tilli, Markku, et al., eds. *Handbook of silicon based MEMS materials and technologies*. William Andrew, Elsevier, 2015.

[13]S. Fatikow, and R. Ulrich. *Microsystem technology and microrobotics*. Springer Science & Business Media, 2013.

[14]P. Dario, et al. *A microrobotic system for colonoscopy* Robotics and Automation, 1997. Proceedings, 1997 IEEE International Conference on. Vol. 2. IEEE, 1997.

is subtle and somewhat deformable. The knowledge and skills of the two environments are not distant from each other, but robotics is characterized by movement in space and by dynamics. It therefore involves a quantum leap in the mechanical and physical sense, because controlling movement in space and interaction with the environment is very complex.

The fascination of robotics lies precisely in this aspect: to create moving systems, in which dynamics and mechanics are the predominant aspects, but which take on a behavior when sensors and controllers are integrated to make this movement fluid, compatible, safe and repeatable according to the robot's operational features. Developing the robot is like building an *alias*, because even if it is neither humanoid nor animaloid, it certainly takes on characteristics that make it similar or "assimilable" to a biological system. As has happened since its origins, robotics has always represented the effort to create an artificial system that does things "in our place". In constructing this, imagination can follow criteria of bio-inspiration or invent new forms and behaviors.

Geography is an essential compass to face our journey in robotics. In fact we cannot understand what robotics is in a universal way, because the social impact and the cultural intertwining are so strong that they deeply influence its evolution. For these reasons the factor of "space" is as equally important as the "time" factor and just as robotics has changed over the years, it depends heavily on the area and geographical location in which it is studied and developed. While industrial automation concerns manufacturing space, and is identical and standardized throughout the world, robotics entering society is necessarily influenced by the context in which it operates and develops.

During my career I have visited Japan many times to carry out experiments or hold classes in laboratories. We had a very close relationship with Waseda University in Tokyo and a joint research lab, also supported by our diplomatic network. In the first years of this century, the support of Italian embassies abroad, especially in Asia, was constant and important for me and my colleagues, and enabled me to understand firsthand how diplomacy today should be not only economic and cultural, but also scientific and technological. Fortunately, at the embassy in Tokyo we met open-minded and creative people who supported us in the opening of the first joint robotics research laboratory between an Italian and a Japanese academic institution in the late 90s. This was a laboratory that later formed a universal model applied by the Italian Ministry of Foreign Affairs. The joint laboratory, of which for some time I was also scientific manager, concerned the creation of a humanoid robot with hands, arms and head able to mimic emotions with gestures and facial expressions. The creation of this "emotional bust" was an important test bench because it enabled us to compare the very nature of the way of expressing emotions, such as joy, satisfaction, enthusiasm or sadness, between the Japanese and the Italian world. Our Japanese colleagues were responsible for the head and the integration, because the robot lived in the Tokyo laboratory, while we Italians were responsible for the hands. We discovered in the field how important geography is and the influence it has on ways of expression and gesture expression. A startling example is how even the simple act of counting numbers with fingers happens differently in Japan and Italy.

It is therefore important to consider this issue in geographical and sociological terms. The main impetus for the development of humanoid robotics in the second half of the 1990s in Japan was linked to the problem of an aging population. It was necessary to find solutions for the personal assistance of non-self-sufficient people which were alternatives to the system of immigrant carers, because Japan is a very closed country and politically contrary to immigration from abroad. Therefore, unlike what happens in Italy and southern Europe, in Japan there is strong resistance to accepting assistance from a foreign person, and instead a propensity to interact with a robot and to appreciate its presence in the home.

Robotics for personal assistance was also an important opportunity for Japanese industry, for giants like Honda, Toyota and Mitsubishi. The 2005 Aichi EXPO was a benchmark for demonstrating the capabilities of humanoid robots produced by research centers in Japanese industries, just as today car manufacturers are competing with their self-driving vehicles. The creation of ever more able humanoid robots has been and still is a test bench for the scientific and technological capabilities of a country or a robotic community. It is worth clarifying that humanoid robotics did not have the rapid and positive evolution that was foreseen in the first years of the century. On the contrary, progress was slow and therefore the interest in the field was also partially weakened, which in any case is still very much alive and arouses attention, for example with international competitions and the related spectacular videos. It must therefore be stressed that creating a robot that walks, grasps and manipulates in collaboration with us, and which is in our midst and in our homes, is an immense problem and still far from being solved, due to a series of technological limitations as yet to be overcome.

Robotics is complex to contextualize as a scientific discipline, but neither is it even identifiable as a set of technologies. Only recently has the international scientific community approved its entry into the field of science, as element of progress with respect to pure engineering. This has been sealed by the publication of prestigious scientific journals, such as the recent *Science Robotics,*[15] whose first issue came out in 2016.

Robotics was originated around the design, manufacture and control of robots, and absorbed and used the knowledge and skills associated with these activities. It was therefore a research area stemming from the integration of scientific and technological knowledge, with a precise purpose: the study and construction of robots. However, it has developed so much that it is now considered a field that passes through engineering towards science and has perfected autonomous and characteristic methodologies and skills.

An interesting and distinctive aspect is represented by the rapidity with which the concept of robot evolves, essentially because as scientific disciplines progress and technologies develop, robotics takes steps forward by applying the results obtained in other fields. In fact, it can be said that robotics depends very much on the progress of other disciplines, such as artificial intelligence and its rapidly expanding areas of machine and deep learning: these produce more effective algorithms and learning methods applied in robots in order to support their decision-making skills and autonomy. Sensors and artificial vision are also becoming increasingly sophisticated and enable robots to have better eyes and therefore to move better and recognize objects in the environment. These examples can be extended to all components or subsystems of a robot, from artificial limbs, to mechanical joints, to batteries, to materials and gradually to smaller but equally important components. Without an improvement of single parts, the integrated system does not improve, but robotic art and engineering have the goal of putting together the different components into a system with superior and emerging characteristics and performance. The robot as a complex system is worth much more than the sum of its parts and the methodologies of integration constitute the value of robotic science. We can say that in this sense, robotics depends on the art of synthesis. While the natural sciences break down, separate and analyze in order to identify the various phenomena, and technologies transform scientific knowledge into tools and manufacturing methods, robotics performs exactly the opposite path. It designs, carries out and controls a system that acquires its capabilities, reaches the required performance and becomes intelligent in the space in which it operates. The appeal of robotics is associated above all with this act of integration which, in order to be scientific and transferable, must lead to repeatability, i.e. the possibility of developing materials and methods that can be described and implemented in other robots.

---

[15]For references, visit the website of Science Robotics http://robotics.sciencemag.org.

A peculiar aspect of the design of robotics is linked to aesthetics, not only in the sense of its external appearance, but above all the combination of beauty and functionality. Aesthetics and functionality are closely linked, and I think that robotics is an excellent example of this close relationship. A robot that responds to its operational features, that of grasping and manipulating an object, with fluid speed and movement, is aesthetically beautiful, first of all because it is functional and performs its exercise in a human-like and therefore pleasant way. Part of the fascination of robotics arises from its bio-inspiration, from its being inspired by living systems, from being gradually more and more similar to what for us is a biological model, sometimes overcoming it in the performance of individual tasks, but reproducing the movement of a natural hand or arm, or of a face or facial expression. Recently, an article by an American biologist in Nature[16] has been published that measures just how much the concept of bio-inspiration or biomimetics has been important in the scientific production of recent years: the numbers are impressive. The author notes that biologists often do not sign articles[17] and therefore are not really involved like other scientists in scientific production. In fact, on the basis of these considerations and the observation of the phenomenon in the field of robotics, we can say that the potential of bio-inspiration has not yet been fully explored and therefore true antidisciplinarity has not yet been reached. There is therefore an unexplored world. I think that biologists still have a lot to teach us and we can still work together to develop revolutionary social robots.

The concept of antidisciplinarity impressed me deeply when it was proposed to the MEDIA Lab at MIT as early as in the 1990s.[18] Interdisciplinary, in fact, means that researchers and scholars of different disciplines work together, each starting from their own knowledge and their own methods, trying to contribute to a collaborative project. The anti-discipline is instead a very courageous leap in quality, because it involves breaking the barriers and intersectoral differences, to build together a common perspective. It is therefore a question of getting out of one's cultural domain and building a new, common one, developing new methods. In some respects biorobotics is

---

[16]Emilie Snell-Rood, *Interdisciplinarity: Bring biologists into biomimetics*, « Nature » 529, 277–278 (21 January 2016) https://doi.org/10.1038/529277a.

[17]They are not therefore involved in the so-called scientific "authorship" of articles.

[18]To see how antidisciplinarity is understood at the MIT MEDIA LAB visit https://www.wired.com/2016/03/mit-media-labs-journal-design-science-radical-new-kind-publication/.

an antidisciplinary field because it involves breaking with the paradigms of the disciplines from which it started and the definition of completely new languages, methods and approaches, that have been consolidated, in books, articles, courses and workshops in its thirty years of scientific and technological history. We must however be very careful because the anti-disciplinary approach is possible only starting from rigorous knowledge and personally I am opposed to an anti-disciplinary teaching from its bases: it is like practicing a sport without knowing the fundamentals. Initially in one's own curriculum one must learn and deepen the methods of a single sector before launching into the borderline areas. I am not surprised that this area of antidiscipline emerged precisely in the North American world, where even training is never univocally within a single field, but on the contrary provides a predominance of programs belonging to a certain area coexisting with some courses also in distant disciplines.

On the contrary, in Europe, and in Italy, training is strictly divided into disciplinary areas, and students are not allowed to go beyond their basic courses, thus discouraging them with unnecessary, bureaucratic and absolutely anachronistic rigidity.

I think that today the most important reform in the educational world regards precisely antidisciplinarity. In fact, it is necessary to inspire more freedom in teaching, and to allow greater audacity in trespassing out of one's own domains above all in research and the organization of knowledge. The academic system divided and separated into scientific and disciplinary sectors penalizes interdisciplinarity and prohibits antidisciplinarity, with the result of curbing the freedom and scientific curiosity of researchers. In addition, the scientific sectors are not open to evolving areas, which are precisely those less explored because they are on the boundaries between one science and another. As a bioengineer and biorobotics researcher I have suffered this typically Italian approach, and I believe that antidisciplinarity is practiced in few places, such as the Sant'Anna School of Advanced Studies and the Italian Institute of Technology.

It is not surprising that robotics is so open to bio-inspiration, which is an embryonic form of antidisciplinarity, and is one of the most beautiful forms of 'contemplation of nature' as it was called from the early years of the twentieth century by a Nobel prize for medicine, Prof. Ramon y Cajal,[19] referring to anatomy and physiology. Building a bio-inspired robot is like contemplating and understanding nature, and to do it we need to understand

---

[19]For a brief biography of Ramon y Cajal visit the website https://www.nobelprize.org/nobel_prizes/medicine/laureates/1906/cajal-bio.html.

robotics, but also biology, and the more the understanding of the two worlds is deepened, the greater the aesthetics and the quality of the robot will be.

The close relationship between robotics and biological systems, and above all between robotics and human beings, is deeply linked to the origin of robotics. In fact, the term robot has been used to identify servitude, the substitution of the human servant, and the task that is performed in place of man. Subsequently, this concept has evolved and changed. Robotics was founded for industry and to replace demanding or repetitive work, but then service robotics expanded, and the robot has entered many specialized environments, and finally in recent years, society itself.

The characteristic that distinguishes the robot from a machine is the intelligence that makes it able to carry out a task autonomously, adapting itself appropriately to changes in the environment in which it finds itself in relation to the task it has to perform. The different components related to the physical part, its body, are the mechanisms, the motors, and the sensors it needs to move and to measure what is happening in the environment in which it is operating. Together with the physical part, referred to as *embodiment*,[20] there are also electronic and IT components that are related to the command of the motors or actuators, the reading of the sensor signals and the processing of data in order to manage the actions of the robot that are defined in the control architecture.

The very brief and qualitative description we have given uses a bio-inspired perspective, with a terminology that also refers directly to the description of a biological, sometimes even anthropomorphic, system. The explanation goes back to the origin of robotics. In having to replace or assist a human worker in carrying out a difficult and repetitive task, such as lifting a weight or screwing a bolt, it was necessary to be inspired in the design by following anthropomorphic standards that led to creating mechanical effectors such as arms, hands or robotic pincers.

The robot is an interesting integrated system in itself, because it represents a challenge for the engineer who designs it and develops it for a specific application. However, it is also an object of study because it is an intelligent system that can be used to interpret neuroscientific or bio-inspired models of motor control, of functioning of apparatuses or organs of living systems, including human ones, in order to assess the reliability and effectiveness of the models themselves. This is why robotics has increasingly attracted the interest of computer scientists and theorists of control theory, eager to experiment

---

[20]M. L. Anderson, *Embodied Cognition: A field guide*, Artificial Intelligence Volume 149, Issue 1, September 2003, pp. 91–130.

with behavior models in robots, and especially neuroscientists and biologists motivated to replicate in real systems their knowledge of the functioning mechanisms of biological systems in order to test them. The robot is therefore characterized not only by intelligence, but also and above all by its behavior relative to the environment, during navigation, exploration and manipulation. Intelligence and behavior make robots very interesting also as training and educational tools, not only to bring students closer to coding, but also as pedagogical methods. In fact, robots develop the interdisciplinary approach, creativity and teamwork because they require more skills to be designed, from mechanics, to information technology, to electronics and behavioral psychology.

In the scientific field, robotics has prompted studies ranging from artificial consciousness, to sensory models, to vision, to motor sensory coordination, to the implementation of two- or four-legged walking, becoming tools for the study of neurophysiology and neuroscience. There are practically endless examples of experimentation in robotic animaloid or humanoid systems, motor control systems and practical implementation of movement. It is clear that success in the introduction of a model of sensory coordination in a robot, which for example succeeds in behaving similarly to living beings and have similar performances in swimming or walking thanks to the bio-inspired hardware or software systems, confirms the consistency of the model, but it is not scientific proof of the model's ability to describe biological reality. Between the validation of the model and its real ability to describe the biological system there is a subtle but important boundary, of which science is fully aware. However, the method of implementing models in systems that mimic the behavior of biological systems is very powerful, because it enables us to study their reliability and to investigate the functioning of the biological system in an effort to replicate it. Cybernetics[21] has traditionally dealt with this area, studying the duplication of the biological system in an artificial system: the hand is one of the most creative examples of robotics that has represented a challenge for many researchers over the years.

I myself have worked extensively on the production of the artificial hand, Cyberhand,[22] and I have personally experienced to what extent in a 'proto-type' task such as grasping, holding and lifting a glass without slipping (so-called *pick and lift*), so many components are involved, like the motor

---

[21]N. Wiener, *Cybernetics, or control and communication in the animal and the machine*, first edition: The MIT Press, Cambridge (MA), 1948; second edition: Wiley, New York, 1961.

[22]For information on CyberHand visit http://news.bbc.co.uk/2/hi/programmes/click_online/3580996. stm.

control, the mechanoreceptors of the skin that have to 'feel' the properties of the object, the vision to guide the movement to the target, and the fingers with their highly dynamic performances. All this is very difficult to summarize in a few words, but it is fascinating to see how the task of pick and lift is one of the most studied in robotics, and represents a real testing ground for comparison and fusion of knowledge between neurophysiology and robotics.[23] I can say, on the basis of my own experiments, and of successes and errors, that without a bio-inspired approach, without understanding how this task works in humans, we could never have developed functional robotic hands capable of imitating our ability to reach, grasp, and lift the object. If we then wish to have the ability to manipulate, everything becomes much more complex. This process of integration and fusion of robotics with neuroscience takes the name of neuro-robotics.[24]

An example of the aim of neuro-robotics is artificial touch. For years the purely engineering approach to the creation of the tactile sensor, i.e. the robotic counterpart of human touch, has been based on the concept of measurement, thinking of touch as a measure of physical quantities such as sliding, position, or pressure. This method is effective in industrial robotics, where the same task must be repeated many times with a robotic clamp that must be fast, robust and repetitive, but if one wants to proceed with the imitation of the human hand, in the complex tasks of manipulation, the pure concept of 'measure' does not work. We need to understand the essence of the sense of touch to reproduce it in a robot that has the ability to grasp and manipulate safely. Touch cannot be compared to a system of measurement, and a classic method of a physicist or engineer cannot be used to describe and reproduce it. We must resort to biology and neurophysiology, and then enter the world of bioengineering and neuro-robotics. Just as vision is not measuring images, and hearing is not measuring sound, touch is not measuring pressure or force. We know that to enter this world of transition between the natural and its artificial counterpart, we need mathematical methods and distinctive and appropriate approaches, such as bioengineering can provide us with.

The reason why the study of robotics attracts interest from several sectors, is that it cannot disregard the most diverse areas of knowledge, and is contaminated and contaminant. To describe what a robot is, we have traveled

---

[23]M. C. Carrozza, et al. *Design of a cybernetic hand for perception and action* Biological cybernetics 95.6 (2006): 629.

[24]For information on the Neurobotics project, visit the website http://cordis.europa.eu/project/rcn/74611_en.html.

through time, in space and through the various scientific disciplines. The study of robotics has changed over the years, and from the world of industrial engineering it has spread and has become increasingly popular even in rigorous scientific and very 'elitist' sectors. I am referring to neuroscience and physiology, where robots have represented training grounds of very versatile experimentation, especially in areas where it is not always possible to carry out direct experiments, as in the case of the functioning of the human brain, motor control and sensory-motor coordination.

At this point it will be necessary to better understand this growth and fertilization of robotics, from industrial robotics to service robotics, up to the present day, the age of social robotics.

3

The Socialization of Robotics

Robots have left the factories and started to "inhabit" other spaces, dedicating themselves to "service" tasks. Places occupied by robots are, among others, the inside of the human body in surgery, interplanetary space and the planets, the underwater world, and nuclear energy plants. The aim of "service" robotics is to teleoperate[1] systems in environments difficult to reach or that are dangerous, which humans should avoid, such as nuclear power plants, minefields in war scenarios or the oceans. In practice, it is therefore a matter of developing robots that are more flexible and suitable to operate in very critical environmental conditions. The most representative images that illustrate service robotics are those of the submarine robot that works on the repair or maintenance of ships, or the robotic arm that performs maintenance in the International Space Station. The particular features of these environments and the relative service tasks include the difficulty of reaching places, danger, microscopic dimensions and hostile or aggressive conditions for humans. The technologies developed for automation and robotics in the production and assembly chain have been adapted and evolved in service robotics to carry out new support tasks, or even the replacement of human operators, under the supervision of experienced and qualified personnel who "guide" the robots.

Unlike industrial robotics, the service version is not aimed at automating a particular task to increase the productivity of a processing or assembly line, but is intended to extend and improve the capabilities of human hands in extreme environmental conditions. Take the example of surgery: the aim is to

---

[1] J. Vertut, et al., *Teleoperation and robotics: applications and technology*. Vol. 3. Springer Science & Business Media, 2013.

© Springer Nature Switzerland AG 2019
M. C. Carrozza, *The Robot and Us*, Biosystems & Biorobotics 20,
https://doi.org/10.1007/978-3-319-97767-6_3

decrease the size of the surgical instrument so as to reduce invasiveness, increase dexterity to navigate within the human body or even reduce the physiological tremor of the surgeon, so assisting the surgeon in those microscopic movements that enable operation with greater safety and precision.

Service robotics has caused a silent revolution, under the sea, in space and in other environments.

One of the most striking events has been the exploration of Mars, where an agile robotic pioneer,[2] able to move over rough terrain, has explored, photographed and analyzed soil samples. Regardless of the not particularly anthropomorphic form of the robot, it has already become an alias, i.e. an agent to carry out the exploration of space in our place. The exploration of Mars represents a great victory for Asimov and for the literature that had predicted robots in space. In fact, just as in past centuries we sent "convicts" to colonize new continents or hostile territories, in the same way today we send robots into space, where we cannot go, because we can specialize them and then sacrifice them. So the exploration of space is an example of service robotics that will probably have a much greater development in the near future.

The most striking revolution, however, is that which has occurred in medicine, where robotics in recent years has completely changed therapies, tools, protocols and also the way to conceive and perform surgical procedures.

I personally experienced the explosion of the surgical robotics industry. At the beginning it was called "computer-assisted surgery", to emphasize the introduction of information technology and image processing in surgery. Then information technology and electronics were followed by the diffusion of authentic support robotics and even replacement of the surgeon's operational activities. Nowadays it is not possible to think of an operating theatre in a hospital, in an advanced country, without the presence of robots. We know that especially in some areas of surgery, such as minimally invasive surgery of the abdomen and urology, the advent of robots has been revolutionary.

Thanks to research work, robotic systems have been perfected to the point that today they add value to the performance and capabilities of the surgeon, assisting in particular tasks such as exploration and operation into the human body avoiding large cuts, so becoming minimally invasive, thanks to the reduction in instrument size. All this has enabled, metaphorically, the creation

---

[2]A. Ellery, *Survey of past rover missions*, *Planetary Rovers*. Springer Berlin Heidelberg, 2016. 59–69.

of microscopic extensions of the surgeon's hands, entering the 'human body without drastic interventions and reducing recovery time for the patient. In other cases it has even made it possible to increase the safety of the procedure by means of augmented reality and sensors distributed in the tools. This is the case in neurosurgery, where vision is improved thanks to the registration, synchronization and recording of the operation field with pre-operative images in the so called 'augmented reality'[3] system where the 3D model of the patient is superimposed on the images collected by the surgical tools.

In just a few years, surgical robotics has passed from research laboratories, to pioneering experiments and eventually to daily clinical practice. As always, the introduction of technology has sparked debate, initially creating problems of acceptance by more traditional surgeons and some resistance to experimentation. Subsequently, thanks also to the generational turnover and the work of enlightened surgeons who have perceived and then promoted the potential of these techniques, robotic surgery has spread and been accredited. Clearly the scientific method and the use of technology are very welcomed when it is actually appropriate and this can only be demonstrated through rigorous clinical testing. The case of robotics in surgery can be considered emblematic of what can happen in other sectors of applications compared to automation, both for the need to adopt an experimental method but also for the crucial theme of the relationship between person and technology and its psychological, sociological and ethical implications.

In the first half of the nineties I personally witnessed the expansion phase of robotic surgery. This originated from the great potential offered by the development of information technology and robotics in order to overcome the limits of manual intervention alone. This transition was also characterized by technical difficulties in the realization of miniaturized surgical instruments, with autonomy, which were controllable and possessed the requirements of safety, biocompatibility, sterilisability and reliability typical of biomedical applications. There have been some areas, such as neurosurgery or vascular surgery, in which technologies have created levels of miniaturization of invasiveness or the recording of preoperative images to guide instruments inside the human body, which have significantly improved existing procedures. On the other hand, there have been other areas of specialization in which robotic surgery has completely revolutionized protocols, leading to completely innovative practices.

---

[3]L. Adams, et al., *Computer-assisted surgery*, IEEE Computer Graphics and Applications 10.3 (1990): 43–51.

This revolution, gradual but inexorable, was embodied by the Da Vinci Robot produced by Intuitive Technology,[4] an American company in Silicon Valley. The robot consists of a rigid structure to which small and agile arms are attached with miniature "hands". It is equipped with multiple degrees of freedom that enhance the dexterity and navigational ability to reach the surgical site by minimizing the opening and thus avoiding invasive cuts. In fact, this highlights the key concept that led to the development of robotic surgery: the realization of microscopic "hands" consisting of micro-locks with built-in tools to operate within the human body, minimizing the surgical wound, and therefore the recovery period and clinical risk compared to general open field surgery. The key elements were, therefore, minimum invasiveness in the human body, the integration with computer science and electronics to enable an increased vision through the pre-operative images that guide the operation, and sensors developed to improve safety and dexterity.

The discontinuity created by the introduction of this robot is evident. For the first time it is the robot that operates in the surgical field inside the patient and the surgeon does not directly touch the operative site, but *teleoperates* at a certain distance by manipulating a specific interface. The surgeon "remote" from the direct operating field is a veritable technical discontinuity similar to that of the fly-by-wire[5] in which the pilot imparts commands for flying the plane electronically, and not through electromechanical systems. In fact, the surgeon is "close" to the operating field, but controls the operation by acting on appropriate "knobs" that drive the so-called end-effector, i.e. the micro hand of the robot actually operating into the patient's body. The Da Vinci—originally intended for cardiac surgery—today has developed in other fields such as that of prostate or abdominal surgery. The robot does not "replace" the surgeon, but is (tele)operated and is under the surgeon's control.

The data tell us that *Intuitive* is a well-established startup in a successful company, which had a reported 2016 income of 670 million dollars. Since 2000 thousands of robots have been sold in the world, but the discussion is still open on the effectiveness of this approach in surgery.[6]

This change was certainly encouraged by the "market", that is, by the will of the patients, who positively exerted their pressure, directing the choice of

---

[4]Surgical, Intuitive, *The da Vinci surgical system* Intuitive Surgical Inc., Sunnyvale, CA, available at: http://www.intuitivesurgical.com (2013).

[5]For a brief description of the *fly by wire* visit the website http://www.aviationcoaching.com/cosa-e-il-fly-by-wire/.

[6]For an idea of current debate and controversial themes see: http://www.healthline.com/health-news/is-da-vinci-robotic-surgery-revolution-or-ripoff-021215#1.

clinics and surgeons to the structures equipped with surgical robotics. It is not entirely positive that the market, especially in the field of medicine, tends to push in the most appropriate direction for the patient, because there is often the influence of a range of factors such as those that cannot be controlled scientifically. For this reason, the technology assessment sector[7] deserves particular attention, in order to study the appropriateness and effectiveness of surgical robotics so as to facilitate the transfer from research to the clinical practice of robotic systems, taking into account the ethical, economic and regulatory standards of public health.

About a year ago I attended a presentation by the general director of a large hospital, who explained in detail the gradual process of accreditation and integration of surgical robotics into clinical practice. I was impressed by the amount of silent work that produced the result beyond mere surgery. This included the management of the quality of materials and equipment, the training of personnel and also the risk management of the responsibility regarding operations.

The progressive application of robotics in medicine has been a scientific challenge that has involved many thousands of researchers all over the world. This has led to the development of biomedical robotics, and the growth of this disciplinary field, incorporating models and methods typical of bio-engineering applications.

From a scientific point of view, service robotics gave a strong impulse to so-called "human-machine interfaces", essential for teleoperating the robot. The interface,[8] the tool that allows the operator to control the robot, is therefore essential. The development of the interface area has been and will continue to be fundamental because the quality and the ability to adequately control the actions of the robots depend on them. Interfaces can be of various kinds and nature: software interfaces installed on computers and displays, with knobs or joysticks for manual control of a traditional type, or even wearable interface, such as exoskeletons or sensorized suits that read and interpret the movements of the subject so as to translate them into commands for the robots. In fact, the field of interfaces has become one of the most important for robotics, but has developed mainly thanks to its use in video games and interactive systems compatible with normal personal computers.

---

[7]G. Turchetti et al., *Economic evaluation of da Vinci-assisted robotic surgery: a systematic review* Surgical endoscopy 26.3 (2012): 598–606.

[8]P. Arcara, and C. Melchiorri, *Control schemes for teleoperation with time delay: A comparative study*, Robotics and Autonomous systems 38.1 (2002): 49–64.

Interfaces are defined for two fundamental tasks. On the one hand they receive instructions for the movement of the parts of the robot, by reading the commands given on the knobs or on the joysticks. On the other hand they supply feedback to the end user through the display, or through a true "return of force" which provides tactile feedback, reproducing the force exerted by the effectors of the robot. The scientific area of return of force is part of a development that has now assumed a dimension in itself, and which is now a field of robotics proper. It involves the so-called haptic return,[9] concentrated on the sense of force and movement, and aimed at giving feedback to the subject who wears or controls the interface in order to give the direct sensation of the perceptual physical experience of the robot effector in the operating environment. This becomes essential when the user operates in a virtual environment, and through the interface they obtain a perceptual illusion of exploration of the virtual object, generated in reality by means of motors and sensors placed in the interface that stimulate the hands and the fingertips of the user synchronously with the exploration itself. Basically, the haptic interface lets us feel the immaterial and virtual object as if it were real, and gives us back the sensations we would have in touching it directly.

Therefore, service robotics deals not only with the robot that performs the action assigned to it, but also with the control system which, with different degrees of autonomy, allows teleoperation of the task by the human operator.

We have emphasized some concepts that determine the succession of the stages in our journey: the exit of robots from the factories, their entry into specialized environments (such as the hospital or operating theatre), the tools dedicated to performing the particular tasks[10] as extensions of the operator's hands, and interfaces that allow the operator to control or teleoperate the robot's activities. There are many other fundamental components of the robot responsible for its operation, and these are less evident in the use of the robot by a person, such as sensors, motors, the control system and learning algorithms. All these components are similar in the industrial robot and in the service robot, but must be designed and adapted to the tasks and mechanical characteristics that the robot takes on.

What are the social and scientific consequences of this evolution for the integration between man and machine? The progressive entry of the robots in hospitals, in the operating theatre, or in hospital wards for logistics related to the administration of meals or therapies to patients, in submarine

---

[9]V. Nitsch, and F. Berthold, *A meta-analysis of the effects of haptic interfaces on task performance with teleoperati on systems* IEEE transactions on haptics 6.4 (2013): 387–398.

[10]In robotics they are called *end-effectors*.

environments or in recovery and rescue operations in the case of environmental disasters, has caused a substantial progress in the acceptance of robots by people. Robotics in medicine or in space takes on a positive connotation, because it makes it clear to public opinion that the robot is assisting the human operator in a useful and beneficial action and in a reassuring way. This is different from the robot's presence in the manufacturing context, where robotics is seen, especially by trade union organisations, as competing with and substituting jobs.

However, to fully understand the impact of the fourth industrial revolution, it is important to underline how from a sociological point of view the shift from industrial to service robotics represents a progressive *socialization* of robotics, which has led to a first visible impact on people's lives. In fact, not everyone knows that the products they purchase are manufactured and assembled in high industrial automation lines using robots, but people are becoming increasingly aware of their presence in operating theatres or in rehabilitation clinics to assist surgeons, doctors and therapists in their activities.

The example of medicine is therefore in an attempt to envisage the future.

In fact, the central question of the fourth industrial revolution is the change in the relationship between robot and person. We are facing a "revolution" because we are on an extraordinary journey: the "migration" of robots, or rather of robotic technologies, from industrial production to the tertiary sector and to services in general. We are at the height of the evolution of the robot from the intelligent tool in the assembly line to the surgical instrument that works inside our body, or to the explorer on wheels that probes the terrain on Mars, but we know that in the next few days the robot will enter our homes, and will move in our streets with autonomous driving.

For this reason we may very well be frightened by the future, because for the first time robotics is coming to our side to help us to walk, talk or even reason, or is even taking our place to offer services that we first thought of as being completely in our hands, such as cooking food, packaging it or delivering it to those who ordered it.

I think when we get to the point where the robot will also choose the food we like best and deliver it to us at the time we want to eat, we will have reached perhaps the top (or bottom?) of this industrial revolution.

In the near future there is a huge challenge for the human and social sciences, to understand and analyze the transformation we are experiencing. It is a challenge in which, as we have seen, *antidisciplinarity* is crucial, precisely because we *hard* scientists and engineers do not have all the cultural tools to understand and face the humanistic implications of the industrial revolution.

Following an anthropological perspective, robotics serves to overcome man's limits. Initially, the robot was designed by responding to the concept of *servant*, that is alias of the human servant, in order to respond to a strict requirement to replace the human subject in performing a particular task immediately and rapidly, perhaps even exceeding human performance, but only in that particular task. It could be said that the robot was created to overcome the limits of the human subject, in terms of resistance to fatigue and safety, and is designed for the carrying out of demanding and repetitive tasks with extreme accuracy. The specialization of the robot works at the expense of a quality that instead remains purely unique and human: versatility. The robot is a specialized servant, who does not tire, even if subject to wear or tear, and as a machine has potentially greater resistance than the human person, and is cut out and built to meet the needs of a particular task.

Among the so-called technical specifications of a robot for applications in industrial automation there is certainly the overwhelming aspect of the working environment and safety compared to the operation with people present in the working space of the robot. So we come to one of the crucial elements to understand this most recent transition: the ability of robots to be in an environment where people are present is not to be taken for granted or easy to obtain. One thing is to develop an agent robot in its enclosed position, where at best only trained personnel enter, and another is to allow the robot to come out and be in contact with people.

From a mechanical point of view the danger is certainly linked to the size of the robot and its speed of movement, and therefore to its dynamic characteristics. A robot that raises a very heavy and large object must in turn be much heavier and bigger. If the speed of operation has to be high, as required by an assembly line, in welding or painting for example, the power generated by the movement can be controlled, but the risk of losing control and causing damage to the surrounding environment and to people remains high. I like to use the "dinosaur" metaphor of Jurassic Park: until the dinosaur is safely enclosed, it is beautiful and spectacular to see, and there are no problems, but it is important that no one enters its space of movement, or breaks the balance because otherwise the situation can become dangerous. The robot is a powerful "dinosaur" that can also be very agile, and the designer's ability is to make it agile despite the mass and power dynamically deployed during movement. To make the "dinosaur" able to be among people, it is necessary to build it safer and able to control its interaction with the human body and the environment, and lighter, more flexible and more appropriate. The "control and lightening of the dinosaur" are the transformations that industrial robotics is undergoing to become service robotics and subsequently social robotics, posing some challenges regarding the evolution of the robot's body and above all regarding its controller. It is well known that the biomechanics of the human arm make it superior to any robotic arm, precisely because of its ability to adapt not only speed and force, but above all the stiffness of the musculoskeletal system, according to the action to be taken, in order for the arm to interact safely with the surrounding environment and to manipulate objects. The biomechanics of the dinosaur and of the human arm, help us to illustrate the challenges of robotics and give us a future reference point in terms of perspective.

To understand the influence of the operating environment in the evolution of robots, it is necessary to cross the spaces where they (be they industrial, service or social) are used.

The industrial robot is very agile, but robust, confined in an enclosed and specialized environment, and is active in repetitive and programmable tasks, which are sometimes very complex, but focused on the objective of reproducing work and assembly operations cyclically and quickly, with the minimum number of errors possible. The assembly line robot is therefore designed 'on purpose' according to the action to be carried out and operates within islands monitored remotely by ad hoc trained workers.

The reason for the distance is safety. Given that in the automotive field the robot operating on the assembly line has to lift and handle heavy objects and quickly, the dynamics of these movements involve very heavy robots that move rapidly and this can jeopardize the safety of persons. The risks of collisions and accidents that can cause irreparable damage to the worker are high. One can therefore clearly understand why the "dinosaur" remains confined to the "enclosure" and why safety procedures are very strict.

The image of the "dinosaur robot", however, may also be linked to another danger at the center of public debate, that of the destruction of labor. If industrial robotics is the result of efforts to improve productivity and product quality, it is also true that it can be seen as competing with the workers' jobs. The relationship between technological innovation and work has been a crucial theme of all industrial revolutions, also due to the tendency of capital to decrease the number of employees in a given production context as a competitive factor in terms of costs and risks. One of the main reasons for the development of industrial automation has been the reduction in the number of workers at a single assembly station, for reasons of cost, efficiency and quality of production. The robots "have won", in the sense that they have become more widespread, particularly in cases where the operations are elementary, and require very accurate movements repeated at high frequencies, but they "have lost" in complex manipulation operations when they must dynamically adapt to the changes in the objects to be manipulated and in the environment. This can also occur in very advanced production plants of motor vehicles where the presence of human operators prevails especially in the last stage of production. This is the phase dedicated to the final assembly of parts in the passenger compartment, where dexterity, handling and versatility are important. In the first stage of machining segments the production is highly automated, and the workers have a monitoring and supervision role, or of selection and loading of materials. At the current state-of-the-art the biomechanics of the upper limb remains unattainable as regards performance when the operation requires dexterity and versatility. The robot can be the best solution in cases where it is possible to carry out the "specialization" in a particular task, optimizing the design with respect to specifications, but in this case the robot becomes difficult to reconfigure, and therefore not very flexible and adaptable to changes, requiring time and work to be reprogrammed.

The world trend of industrial production is to reduce production costs, lower the number of personnel employed and the risks to their health, and improve the quality and speed of the production of the pieces. These objectives prompt process automation, limiting human intervention to complex manual tasks. Not only is this true in the automotive field. I have seen with my own eyes manufacturing environments in the agro-food field where you can make almost all the lines perfectly automatic except for those in which human sense-motor coordination is unequalled, such as in operations where there is the need to recognize the stage of ripeness of the fruit, the consistency of a mozzarella, or the variety and size of an apple during the composition of fruit boxes from multiple cropping.

There are those who maintain that this evolution towards a totally automatic factory represents an inexorable trend, which eliminates operational positions that are simple, repetitive and weary and creates jobs with a higher level of professionalization, even if in much lower numbers. The relationship between technology and work has marked all industrial revolutions, and there are those who talk about the *end of work*[11] and the need to introduce a guaranteed minimum income to make up for the lack of jobs in the machine society.[12]

In fact, industrial robotics has changed production processes, modifying the factory environment itself, and making the work completely different. In

[11]J. Rifkin, The End of Work: the Decline of the Global Labor Force and the Dawn of the Post-Market Era , 1995, Putnam Publishing Group.

   T. Pearson, *The End of Jobs: Money, Meaning and Freedom Without the 9-to-5*, Lioncrest Pub. 2015.

[12]Daniel Stanley, *Elon Musk: Automation Will Force Universal Basic Income*, Geek, May, 29, 2017.

parallel with the development of microelectronics and mechatronics, it has constituted one of the technologies characterizing the technological paradigm of the third industrial revolution. This change in work, if on the one hand it has brought about a reduction in the work that is arduous and dangerous for the workers, and "the distancing of the worker from the workpiece", in order to preserve his health more effectively, has certainly decreased the number of workers dedicated to a single production line.

It is also true that often the solution chosen by the ruling class to improve productivity has been to decentralize the work, trying to lower its cost, rather than investing in technology and improving production in terms of quality and efficiency.

The theme of using robotics and industrial automation at the factory is associated on the one hand with the fear of losing jobs, and on the other with the hope that work is made safer and better, and finally on the idea that the robot can allow human limits to be eliminated indefinitely. In reality, robots are also subject to the problem of wear, the difficulty of reconfiguration, machine downtime, and maintenance. We must fight the illusion that machines are immortal and free of defects, because in a sense this also feeds the frustration and the sense of powerlessness when faced by them.

The emerging industrial revolution, a process of technological change that was already largely underway, has caught the attention of media and politics only recently. A series of prospective studies by researchers[13] have come out, and later surveys produced by various agencies and international institutions. These have highlighted the relationship between artificial intelligence, automation and work, also in the service world.[14] According to Benedikt and Osborne, at least 45% of jobs in the United States are at risk of being canceled due to the introduction of artificial intelligence and automation. This study has prompted a series of analyzes and discussions that are still ongoing and almost completely pessimistic.

Few, however, have fully grasped the further implications of the progressive phenomenon that corresponds to the escape of robotics from the domain of manufacturing production, and its gradual entry into civil society, into our

---

[13]C. Benedikt, M. Osborne, *The future of Employment*, Working Paper, Oxford martin School, Univ. Of Oxford, Sept. 17, 2013 http://www.oxfordmartin.ox.ac.uk/downloads/academic/future-of-employment.pdf.

[14]World Economic Forum https://www.weforum.org/agenda/2016/12/stephen-hawking-this-will-be-the-impact-of-automation-and-ai-on-jobs.

O. Williams-Grut *Robots will steal your job*, Business Insider UK, Feb 15, 2016 http://uk.businessinsider.com/robots-will-steal-your-job-citi-ai-increase-unemployment-inequality-2016-2.

homes, to carry out domestic tasks, or in our streets, where it will help us to drive vehicles in the world of logistics and personal mobility.

Metaphorically, the phenomenon of socialization of robotics can be thought of as a "cycle": we start with robotics and industrial automation, move on to service robotics, and then to social robotics. Finally, there will be the wearable robot, for support or functional replacement, consisting, for example, of prostheses or exoskeletons, which we shall address in later chapters.

The *social* robot accompanies us, entertains us and finally drives our cars in our place, circulating on our streets, but also living in our homes, to deal with household chores.

The extreme point of the cycle, not yet fully achieved but very much underway, is the "implantable" robotics in our body, the intelligent and biomimetic endo-prosthesis, which replaces all or partially the organs and parts of the human body, entering into contact with our inner world through neural interfaces. Following this approach, neuro-robotics, therefore the integration of robotics with our brain, and bionics,[15] which produces implantable parts and neural interfaces, represent an extreme stage, in which robotics does not replace us, but enters into symbiosis[16] with us.

In short, therefore, robotics represents an opportunity, or a threat, according to one's point of view, for the replacement and the end of work, as we understand it, and will probably modify the quantity and the variety of occupations that we can carry out. However, in a different way it will be able to come into contact and change human nature through bionics and neuro-robotics because it can also alter or enhance human capabilities.

The socialization of robotics involves other areas, such as economic, managerial and civil law aspects, due to the impact it may have on society. Robotics enters the dynamics of society, and the remodeling of jobs is only part of this "great transformation".

In this regard, we have seen that in the factory it has led to a transformation in the way of working, and has affected productivity, the quantity and quality of jobs. We have already described how robotics has affected many other specialized areas, for example surgery, where it has changed surgical protocols, instrumentation and consequently impacted on the training programs for medical and paramedical personnel dedicated to surgical operations.

---

[15]For more on the frontiers of bionics, visit the website: http://journal.frontiersin.org/journal/all/section/bionics-and-biomimetics.

[16]A well-known theologian, my good friend, and scholar of technoethics, Josè Maria Galvàn, criticized my use of the term *symbiosis* because it refers to two living beings—the robot is artificial.

Returning to the "socialization" of robotics, it is the cognitive relationship between the robot and those who use it that plays a prevalent role. This is for two reasons. The first is that we will delegate to the service or domestic robot also a part of our cognitive activities and not just physical tasks, as occurred at the time of the entry of automation into the factory. The second concerns social interaction with the user. We cannot predict what will be triggered by the exchange between user and robot with a level of linguistic and cognitive complexity as high as the one we have for example with the smartphone. It is as if the smartphone has taken on corporeity and helped us to perform physical work, but in doing so, it interprets our profile and our orientations.

These are two very important areas that deserve greater in-depth observation, each in a separate way, even if it is the combination of the two qualities, cognitive and physical, which constitutes the strength of the "social" robot.

The delegation of cognitive activities takes place when we use our computer or our smartphone, and we more or less consciously use the tools to support productivity (so they are called in Microsoft and Apple), and they help us to make better presentations, to organize our agenda, and to book a holiday as we prefer, i.e. by following our user preferences. The impact of robotics is expressed in my opinion by combining these potentialities, already offered by the computer, with those of physical presence, the ability to generate movement and perform tasks, providing new possibilities for interaction.

So we will delegate cognitive and physical activities to the robot that will potentially become also an alias that can act in other environments on our behalf or in our place, interpreting our desires and our personality. Licklider,[17] one of the fathers of current computer science, had already foreseen everything in the 1960s. In fact, in a famous article he described how we would arrive at a symbiotic relationship with our computer. In this regard, patents have already been filed to protect the idea of creating robotic agents with our personality, as could be inferred from our web activities. This happens for example when we use Google software[18] and we are "profiled" through a process of indexing and processing of data that we exchange with websites. The profile is generally created by Google and sold in the form of fundamental information for commercial and advertising purposes. We should always think that when we type a question into the search engine we

---

[17]J. C. R. Licklider, *Man-computer symbiosis* IRE Transactions on Human Factors in Electronics, volume HFE-1, pages 4–11, March 1960.

[18]K. Darling, *Why Google's Robot Personality Patent Is Not Good for Robotics* IEEE Spectrum, April 2015, http://spectrum.ieee.org/automaton/robotics/robotics-software/why-googles-robot-personality-patent-is-not-good-for-robotics.

are giving away information about ourselves. In the future, I imagine that our profile can be expanded with more and more information on our habits, including our corporeity. There will be not only personal physical data such as weight, height or body mass indexes, but also how much we walk, train, and what we eat. Our smartphone collects an increasing amount of information about us. One day it will be able to transfer the data to an algorithm that builds an alias of our personality to be translated into a physical agent that replaces us, like our avatar. In this way we would have social networks of physical agents, not just virtual profiles. The question we ask ourselves is— will there be social robot networks?

4

Our Friend the Robot

Whenever I think of the future of robotics, what comes to mind is the image of a lady in her comfortable home in her unspecified city. She is quietly preparing a cake for a dinner that is coming up soon and is mixing the ingredients according to the recipe. This is not an unusual scene—rather it is both traditional and universal. It can take place in different parts of the planet, regardless of places and cultures. If we approach and observe well, however, we discover that the lady is not following a recipe book as my grandmother would have done, or a television program, as my mother would do today, or even using a smartphone with a trendy app, as my daughter normally does. On the contrary, the lady is talking with a kind of one-eyed table-top robotic lamp, whose head swivels, responds and follows the movements of her hands, measures the quantities of the ingredients, and corrects or suggests the actions to be carried out—how to mix, how to add more. It even praises or softly scolds if too many utensils are dirtied.

When family members enter the room, the table robot turns to them, recognizes faces and takes pictures, or reminds them of appointments. In its aesthetic appearance the robot is anthropomorphic, but not too much. It assists operations as required, and suggests without being too invasive. It finds recipes on the Internet, measures and remembers, responds to commands, stores information and greets when someone comes into the room. It is a new generation of robots, designed for wide consumption, and provides a friendly and pleasant personal assistance, based on dialogue with the user that can also turn into the administration of cognitive therapy, if necessary.

© Springer Nature Switzerland AG 2019
M. C. Carrozza, *The Robot and Us*, Biosystems & Biorobotics 20,
https://doi.org/10.1007/978-3-319-97767-6_4

I saw on the Internet some promotional videos of Jibo, a social robot of this kind, which originated in an American startup company.[1]

I know little about this robot, and I partially used my imagination to describe it, but I know that the startup intends to launch the social robot onto the market, responding to the need, perhaps already perceived or generated by the robot's own advertising, for a lady-in-waiting, a friendly assistant who interacts, obeys orders and above all is sentient with respect to our orders and knows us and knows what we want, just like a butler. Basically, the table robot knows where its place is.

Will we ever reach this level of sophistication and perfection with consumer robotics? I do not know, but I certainly know that many are trying and that it is important to follow these attempts.

This type of robot is different from what we have seen in previous chapters because it has to do with entertainment or public relations robotics aimed at a wide audience. There are already several examples of commercial robots available on the market, and these are pet robots, therapeutic robots for cognitive rehabilitation, or even educational robots to be used at school or in training programs.

In many cases these robots take on an anthropomorphic form and tend to be small, especially to be easily transportable and used by children and the elderly. The problem of the cultural integration of robotics influences these applications in which the robot tends not to "frighten". For example, it takes the form of a pet, or a toy. The goal is to make itself acceptable, pleasant and reassuring. In fact, the image of the lady who is cooking a cake is reassuring. It represents robotics as an industrial and commercial opportunity, to enter our homes, our intimacy and to help us.

There have been some resounding failures of this approach, like the Sony robot dog,[2] but there are also several positive examples, for example in the field of educational robots, which are very popular in Italy.

Educational robotics is very successful because it can be a method of teaching science and technology, such as computer science, mechanics, physics, electronics and even behavioral psychology, but it stimulates creativity, groupwork and the antidisciplinary approach.

I do not think we will have a substitute robot that will go into the classroom and replace the teacher in giving lessons. However, but I do think we will have more and more robotic platforms that will enable students to carry

---

[1] Jibo is a social robot—for more information, see the website https://www.jibo.com.

[2] Aibo is a robot produced by Sony, for more information on its story, see http://www.imperialbulldog.com/2015/06/23/il-destino-di-aibo-primo-cane-robot/.

out exercises in the laboratory with practical applications of the program subjects, integrating them creatively according to their ideas.

Precisely because of the ability to stimulate problem solving, imagination and the antidisciplinary approach, I am in favor of a systematic introduction of robotics into school syllabuses,[3] with practical courses of experimental production of prototypes and devices that stimulate students' imagination and creativity. Therefore, social robotics offers a wide range of possibilities, even for applications in the medical field, where it is always necessary to start with rigorous experimentation protocols, as is happening in Japan where robots have been developed for a new type of pet therapy with the elderly in rest homes.[4] Once again, Japan is in first place in this field, probably because of its positive cultural predisposition to be helped by a machine rather than by a person, but also due to the need to find social solutions to the aging of the population and the need to develop new welfare systems.

What is therefore the robotics market, exactly? What will it be like in the future? For now, robotics has not reached the production and sales volumes of consumer electronics[5] and despite the market forecasts being optimistic, we do not yet know if any of the products we have described will be successful and reach our homes. In fact, the acceptance of a robot for entertainment or personal assistance involves the intimate sphere and therefore is not easily predictable or foreseeable with market analysis.

Companies and investors who are interested in the diffusion of robotics in society are aware that the assimilation of the robot into the cultural context is essential. In any case, whether one is thinking of a use for services in medicine or personal assistance, or intends to sell robots to keep people company, socialization is essential for acceptance, and, if possible, a certain appreciation of the robot by those that have to buy it and use it in everyday life.

This is why research also deals with these issues, as in the case (cited in Chap. 1) of the humanoid robot which, like an actor, interprets emotions through an expression of the face associated with gestures of the arms and hands. We have already stressed, with reference to our experience in Japan, that interpretation is not universal but strongly depends on cultural context.

---

[3]At the *Scuola Superiore Sant'Anna* there is a group headed by Prof. Paolo Dario, which develops experimental educational robotics projects together with Tuscan schools and in cooperation with school teachers.

[4]Paro is a therapeutic robot, whose characteristics are described on the website http://www.parorobots.com.

[5]Data on the market of social robotics indicate that both demand and client satisfaction is increasing. See Reports of the International Federation of Robotics, https://ifr.org http://bib26.pusc.it/teo/p_galvan/technotehics%20ieee%20tradotto.pdf.

In the same way, in the design of the robot it is not possible to disregard knowledge of the social context in which it operates.

In this regard it is important to emphasize that the main difference between an industrial (or service) and a social robot consists precisely in the fact that the former is a professional tool in the workplace, used by trained people, and can be designed in a universal way. On the other hand, the social robot is used by non-specialized people, who interact with the robot without specific training.

From the point of view of the project objectives, two different ambitions for social robotics coexist. The first is aimed at developing an instrument that performs a service task in our place with very rigid specifications, while the second stems from the idea of achieving an alias that maintains a cognitive relationship with us, in a programmable and versatile way.

Some examples may help us understand the differences. For example, there are robots designed to help a disabled or elderly person to do household chores and so-called daily living activities that they would not be able to accomplish on their own. For paralyzed people who have suffered injuries to the central nervous system, or suffer from neurodegenerative diseases, it is very important to have robots able to restore functions related to mobility and manipulation because this enables a partial recovery of autonomy.

In everyday life we move, we get out of bed, we wash, we dress, we prepare and eat our food, and we connect to the computer just to communicate. Disability is not a characteristic of the person, but it is defined precisely in relation to the ability or not to carry out these tasks which, if reconquered thanks to the help of technology, represent a fundamental goal for the quality of life. This list includes mainly physical motor activities, based on mobility, the reaching for objects or within environments, gripping and manipulation. Robotics adds greater strength, intelligence and autonomy to technological aids designed for personal assistance and through appropriate interfaces[6] that are adaptable to the person. This offers the possibility of naturally controlling the instruments by integrating them with one's body and one's motor intention.

When I consider the robotics sector for disabled people who are made more independent thanks to technology, I am reminded of the many people who have used our devices in the laboratory. I recall their stories, their desire to contribute to the research due to the consequences that it can have for all their community, and their smiles in front of the small progress that they get thanks to our joint efforts. This represents a daily stimulus that prompts us to constantly improve robotics for personal assistance. A better robot can make

---

[6]The field of interfaces for controlling external devices, which interpret neural signals, is normally indicated as BCI (Brain to Computer Interfaces).

the difference between standing up on one's own or sitting all day in a wheelchair, being able to pick up a glass or asking for help. With the strength of many small steps this can change the perspective of a disabled person's day, giving back free time to work, socialize or simply think freely.

Regarding the subject of personal assistance, one can imagine autonomous, external or wearable robots, as we will discuss in the next chapter. The autonomous robots for domestic use move thanks to a navigation system in a known environment with acquired internal maps. They can have one or more manipulators to reach objects and grasp them, and then to take them to the person who commands them. They have an anthropomorphic form, mainly because they have to move in an environment designed for humans and therefore the greater they resemble physical persons, the more they conform to the environment and the functions they have to perform.

As far as navigation and manipulation aspects are concerned, the artificial intelligence and the guidance and learning system are very similar in all robots, whether they are for service, personal assistance or entertainment. The main difficulties are related to safety in the relationship with the human user, and above all in the universality and unpredictability of the environment. We have understood that the logical jumps are enormous—that is from the imprisoned dinosaur robot, to the collaborative robot with a trained and professional worker, up to the robot at home with a weak and untrained end-user. The fourth industrial revolution involves precisely this transformation of the robot, with the related scientific and technological challenges connected to this sea change.

When the robot becomes a consumer product, aesthetics and industrial design become very important. For example, an anthropomorphic configuration can be an advantage in terms of its acceptance by persons and to ensure greater operational features in relation to the environment and to the actions that the robot has to carry out. However, it is not obvious that this is the best solution or the one that will have more success on the market. Some trends go in the opposite direction. We can think of domotics (home automation), where the smart and functional devices are created inside the home, as an evolution of home appliances and systems, and are controlled through an infrastructure immersed within the architecture of the house. This is a progressive robotization of household appliances, without changing their use or appearance, as happens in an emblematic example represented by the Roomba vacuum cleaner from IRobot.[7] Roomba has been sold in its

---

[7]Roomba is a robot produced by IRobot http://www.irobot.com.

thousands and has brought robotic technology into the home with the appearance of an advanced vacuum cleaner, which has not changed the main function of the appliance, but has increased its autonomy. This is how a transformation of the home takes place with increasingly independent and intelligent appliances without the presence of an anthropomorphic humanoid robot. What is very important in this case is the control interface of the domotic infrastructure that must make the command of the various functions for the householder simple and fun. This is to avoid generating the rejection caused by the sense of frustration when faced with over-complicated technology, which simply provokes discouragement.

We still do not know which of the two design architectures will be rewarded most by the market, that of the anthropomorphic servant robot or that of technology immersed within the home and its furnishings. They both represent fascinating industrial challenges that reach beyond robotics and cross over into design, architecture, sociology, psychology, and encroach on more purely humanistic and cultural issues, especially if it is a matter of understanding which system will be more easily accepted by people. The social robotics sector would seem to be the choice in terms of the tradition of Italian industrial design, and I hope that my country will be able to seize the industrial opportunity it represents.

The experimentation methodology, together with the launch onto the market of innovative products with different shades of relationship between use, specifications, and aesthetics, will tell us which system will be more successful and in which areas of the planet. From comparative studies,[8] we know that different cultures, either Asian, for example that of Japan, or that of North America, have differing degrees of acceptance of a domestic robot. There are also differences between Northern and Southern Europe, where traditionally we tend to prefer the help of a family relative to the presence of a robot. On the other hand, in Japan the propensity to use and accept robots at home is very high and, probably thanks to the Manga comics, there is a widespread tendency to see them very much as heroes rather than villains.

In all these examples, the common key is that robotics is intended to overcome man's limitations. It creates instruments with differing degrees of intelligence and autonomy that finalize human intentions, enabling persons to perform tasks otherwise impossible without technology.

---

[8]Kitano, Naho *A comparative analysis: Social acceptance of robots between the West and Japan* EURON Atelier on Roboethics (2006).

Heerink, Marcel, et al. *The influence of social presence on acceptance of a companion robot by older people*, « Journal of Physical Agents » 2.2 (2008): 33–40.

This nature is often at the center of mass media exaggeration and anti-scientific literary or cinematographic interpretations, which can lead people to the illusion that the machine is infinitely powerful.

The frustration when faced with a presumed infinite power of machines also prompts the fear of abuse of power and an obscure domination of technology and of those who are able to design it. This is why it is important to emphasize what is apparently obvious, but which is easily forgotten. Human limits are never canceled by research but only 'moved forward', thanks to a series of systems, also finite and subject to imperfections, however advanced and well-designed they might be. On the other hand, we are also aware that robotics can be used incorrectly or maliciously, regardless of the reasons for which it is designed, so we must discuss the possible uses of robots, and the responsibility of the engineer. This theme is at the intersection of science and ethics and is dealt with in a discipline that we introduced already in Chap. 3, technoethics,[9] which helps us to reflect on the purposes of technology, according to different approaches and different possible positions.

It is worth pointing out that one of the most flourishing areas of robotics is the military, or defense sector, and that every engineer poses the problem of the final application of technologies to which he is contributing according to his own morals, but technoethics must give universal tools in order to interpret these issues.

I believe that the purpose of robotics is love for our neighbor and brotherhood, which can be embodied with some examples. These include improving the quality of life of a sick or disabled person, supporting an operator's demanding work in a factory, or rescue robotics to avoid the risk of contamination after a nuclear accident by replacing the personnel with an explorer robot on the ground. In these cases, the technology is used to overcome a limit and therefore makes for greater safety, and to reduce damage or to help people. In the European Union, the social innovation sector indicated in the Horizon2020[10] program sums up well this positive trend in research in terms of benefits for people.

The motivation of social innovation is shared by many colleagues, especially Europeans, but there are other perspectives and other very different

---

[9]J. M. Galvan, On Technoethics, http://bib26.pusc.it/teo/p_galvan/technotehics%20ieee%20tradotto.pdf.

[10]Horizon2020 is the research program of the European Union, https://ec.europa.eu/programmes/horizon2020/.

positions. This is why the discussion in technoethics must always be alive and open to contributions from scientists all over the world.[11]

From the point of view of the final products we have described cases of robots working with us under the same roof, or in our place in a given environment, interpreting or identifying in advance our desires, or simply obeying our commands, in a sentient way. I tend to emphasize the term "socialization" because it is a process of entry of the robot into civil society. Now we are used to carrying out tasks that will soon be performed by some robot who can drive our car, do our chores, take home the food we order on the Internet, or keep us company reminding us of our daily tasks. All this is possible only with the combination of various factors in the robot:

(1) physical strength, movement and manipulative abilities;
(2) an artificial intelligence providing some cognitive functions in learning new tasks and in interacting with human beings.

The final outcome is that the robot is, autonomous, adaptive and able to react according to user needs and changes in the environment.

It is a challenge that research is accepting, but with much still to be done, above all to ensure that the control of interaction between the robot and the human subject takes place safely and without damage or injury. Moral laws will also need to be by robot behavior, and legal procedures must be defined to establish the responsibility of the robot as regards property and the persons with whom it comes into contact. The movement that supports the entry into the car market with autonomous driving is precisely aimed at creating the physical, but also regulatory, moral and legal infrastructures that enable such a secure entry of artificial intelligence, automation and robotics into society, in our midst. We do not know yet when it will happen, but we know that sooner or later it will happen. This transformation will put us once again before the theme of limitation. There may very well be unforeseen accidents and events that will demonstrate the limits of the robots, even to those who thought they had none, and they will frighten those who will have to insure the robots and give permission for their circulation. Furthermore, there are emerging and very interesting issues concerning the sustainability of the production of robots and the materials they are made of, in terms of developing a circular economy. Today we know that in the automotive field we are

---

[11]Technoethics deals, and in the future will increasingly deal with the relations between robotic systems and the military environment. In this book we do not deal with the use of robots in warfare and their relations with defense and weapons, but it is worth highlighting that there is constant debate on the topic, which must be made known to the general public.

moving towards the production of vehicles that can be completely dismantled and recycled, and it is unthinkable that this issue is not also be addressed in the field of consumer robotics. The ethical, legal, design and environmental sustainability issues will slow down the process, but the advance is now unstoppable. Some form of robot more or less resembling us, humanoid or animaloid, or masked as a vacuum cleaner or intelligent vehicle, will be able to live with us, in our homes and on our streets.

For this reason I think that we should not passively put up with this phenomenon with bewilderment, but rather try to understand it, foresee the consequences and prepare ourselves to face it. I think that our young people should learn the grammar and language of this change, not so much with a digital education as simple users, but with the study of methods. This means the coding[12] that enables them, regardless of other subjects in which they specialize, to be programmers and protagonists of change, not merely passive consumers of applications.

Joseph Licklider (Chap. 3) predicted the "symbiosis" between man and computer, which today has led to the sharing of objectives, so producing "interdependence". This is an extraordinary foretaste of the potential of the computer based on the adoption in computer science of a purely biological concept, symbiosis, which involves a close interaction between two biological subjects for mutual survival. Even if the concept of symbiosis applied to a natural agent with an artificial one is not completely consistent with its meaning, it can be very useful as a metaphor to understand how the computer or smartphone are becoming symbiotically present in our life. This is to the extent that we use them to communicate with others in a sometimes spasmodic and excessive way, and even if we are aware of not having total control, we cannot do without it, because it is now an instrument that "shares our daily goals and allows us to reach them" as Licklider had predicted. We do not yet know how robotics can reach these levels of intimate communication and cognitive connection with the user, but we expect this theme to become predominant.

In Chap. 1, we mentioned that in industrial revolutions there is a strong tendency for capital to finance the construction of machines that perform tasks in the place of persons. In today's phase we are witnessing the exasperation of this trend, which may represent the so-called "end of work" but also the development of another human aspiration, that of building a twin, an

---

[12]Carrozza et al. *Istituzione dell'insegnamento di principi di informatica nella scuola primaria e secondaria* (2633), (In Italian) Bill presented at the Italian Chamber of Deputies, 15 Sept. 2014, to see the text consult: http://www.camera.it/leg17/126?tab=&leg=17&idDocumento=2633&sede=&tipo=.

alias that represents us. It is possible to imagine that the development of robotics, combined with the technology of materials and artificial intelligence, makes it possible to construct replicas that closely resemble us from a behavioral and aesthetic point of view and represent us in society. We are not talking about science fiction, but about research projects in progress, especially in Japan and Korea.[13] These are contributing to the creation of alter egos with faces and expressions totally resembling the creators of the robots, as if they were their personal robotic twins. It is a field of research that goes beyond the boundaries with the social sciences, involving 'custom', and poses sociological questions. To non-experts, these experimental methods can generate a mixed feeling between enjoyment, dread, repulsion and fear for the future.

Let us try to think of the substitute robot entering the classroom in place of the teacher, who is teleoperating it from home, or politicians who flood their electoral territory with their own doubles on public occasions. So we can imagine a world in which we have a network of robots that interact with each other in our place.

In fact, it is a controversial and debatable subject, which with its extreme consequences leads on the one hand to building robotic aliases proper, which tend to resemble physical persons perfectly, and to imitate their expressions or gestures, or on the other hand, prompting the robotic evolution of dolls for sexual entertainment, with "replicating" robotic systems resembling a beautiful, young and attractive person. I will not dwell on this type of application, which I personally see as degeneration, but I fear that the combination of an anthropomorphic robot, similar to a person, with an appropriately programmed artificial intelligence system, leads to potentially dangerous mechanisms of dependence and attachment. A film that envisages this concept is *Her* (2013). *Her* tells the story of the extreme attachment of a weak person to an imaginary 'operating system' with artificial intelligence. The operating system, simply by voice and the ability to interact, in a more or less humanized way, manages to obtain a spasmodic and all-encompassing attention. We could call it a 'falling in love' if it were not somewhat risky to use this term for a man's feeling towards a machine.

Without reaching the extreme case of *Her*, we can think of small, manageable robots, for example to be kept on the table, which are nice to look at, and highly interactive. With their pleasant expression, their big eyes, and the

---

[13]E. Guizzo, Hiroshi Ishiguro, *The man who made a copy of himself*, IEEE Spectrum, 23 Apr. 2010, http://spectrum.ieee.org/robotics/humanoids/hiroshi-ishiguro-the-man-who-made-a-copy-of-himself.

gestures of their body, they manage to express very simple emotions, they follow the operator in their movements, speaking and managing to interact.

The enabling technologies supporting such a product, like those of material production, or systems for voice interpretation,[14] will favor the interaction and reaction capacity of these robots, and make them potentially usable for delivering cognitive rehabilitation therapies based on games with the user.

I believe that these robots represent the maximum expression of social robotics, because they establish a connection, and they also interact culturally with the user. We have defined them as pet or entertainment robots, but in perspective, as we mentioned in Chap. 3, we know that they have new potential as they may be able to take on the personality of their owner, and interact on the web by taking on that profile and becoming proper avatars. As we have mentioned, if more or less physical or virtual subjects acquire the personality and identity of a user, who delegates to them their interaction with other real, virtual or alias subjects, a social robot network can be created. In this way a physical network of agents is created that interacts socially, giving rise to exaggerations such as those in which a person survives himself by creating his own avatar which is then freed and evolves without control.[15]

In the fourth industrial revolution, the robot network also has a certain strength linked to the cloud.[16] The cloud is a powerful tool, able to create an accessible virtual environment where memories, libraries, programs, and the data acquired from the perceptual experiences of the robots are stored interactively. This creates a form of transferable collective intelligence that is accessible by other robots. This speeds up learning and training time, and therefore the autonomy of the robots. From the point of view of managing a set of robots that cooperate in a given environment, the cloud can enable experimentation with innovative architectures, thanks to the principle of collectivization of intelligence, introducing greater flexibility and operating possibilities. The determining parameter for the cloud is the wireless network and its connection speed. The higher this is, the greater the transmission of high volume data, making real-time control possible without latency,

---

[14]In this regard, a good example is Siri, the iPhone voice assistant.

[15]In the USA there is discussion of investment in robots whose minds it is possible to clone. For further details, see the Bloomberg TV interview, available on the website https://www.bloomberg.com/news/videos/2015-02-04/see-future-of-artificial-intelligence-in-mind-clone-robot?hootPostID=eca0a4210f3d2633d5fa0d2567c71b6c.

[16]Kehoe, Ben, et al. *A survey of research on cloud robotics and automation* IEEE Transactions on Automation Science and Engineering 12.2 (2015): 398–409.

transferring information at ultra-rapid speeds. For this reason, new telecommunication systems are being developed that take the name of 5G.[17]

In the future there will be many more objects than people connected to the Web,[18] which will collect data continuously on their operation, the environment in which they operate and on the activities of their users. Therefore, the cloud will be the accessible tool for organizing databases, and in this perspective, an enabling technology will be made up of Big Data analysis tools capable of extracting sensitive information for the application.

On the other hand, one of the greatest challenges is to ensure data and cloud security.[19] In fact, the greater the connectivity of the robot in the cloud, the higher the IT risks, and consequently the need to make sure there is appropriate control in order to avoid external attacks for malicious reasons. All the governments of the advanced countries are investing in cybersecurity, which in this scenario plays a fundamental role in defense. Therefore, 5G telecommunications technologies, big data[20] and cybersecurity will be enabling technologies in the fourth industrial revolution, together with robotics. In our snapshots of the future, in fact, the theme of security is very present. We can clearly imagine the consequences, with varying impacts, of intrusion into the robotic system that helps the lady to prepare the cake or controls the robotic alias of the politician.

---

[17]For an idea of what 5G is and associated products, consult http://www.techworld.com/personal-tech/what-is-5g-everything-you-need-know-about-5g-3634921/.

[18]IoT/Internet of Things—it is foreseen that in 2020 there will be about 50 billion web-connected objects, many more than the total population of the earth. Coverage today reaches 76% of the most advanced countries, according to the International Telecommunication Union.

[19]P. W. Singer and A. Friedman. *Cybersecurity: What Everyone Needs to Know.* Oxford University Press, 2014.

[20]Chen, Min, Shiwen Mao, Yunhao Liu, *Big data: A survey*, Mobile Networks and Applications 19.2 (2014): 171–209.

5

The Robot Inside Us

One of the most reproduced robotics videos on You Tube shows a Quadruped animaloid robot, similar to a mule, capable of running uphill in the woods on rough terrain and avoiding obstacles, carrying a load on its back. It is a robot with a lot of natural-looking features that represents the brilliant result of an American startup, Boston Dynamics,[1] financed by the United States Defense Advanced Research Projects Agency (DARPA[2]). The project was devised for military applications, where in theory the ability to use impervious routes carrying heavy objects could be of great help, and certainly will have industrial repercussions in terms of know-how and patents for its production.

The real applicability of these quadruped robots is however markedly controversial. So much so that the fate of Boston Dynamics, bought and subsequently sold by Google, is indicative of how long the market forecasts and their strategies are wavering in time. However, the scientific challenge associated with their production is relevant, because designing and making a robot emulator based on an animal model involves the creation of high performance systems equipped with very sophisticated control algorithms.

Within the context of scientific research, animaloids represent an opportunity for scientists to create artificial replicas of natural counterparts, which are useful for experimenting with models of biomechanical movement or motor control developed in the neuroscientific field. This was in the spirit of

---

[1]Visit the Boston Dynamics website http://www.bostondynamics.com.
[2]DARPA (Defense Advanced Research Projects Agency) is the Research Agency of the American Department of Defense. For more information, visit the website http://www.darpa.mil.

© Springer Nature Switzerland AG 2019
M. C. Carrozza, *The Robot and Us*, Biosystems & Biorobotics 20,
https://doi.org/10.1007/978-3-319-97767-6_5

'cybernetics', which is not new (in fact it dates to the early 1900s), but today it is supported by very advanced technologies compared to the past.

Bio-inspired robotics has produced various models of humanoid or animaloid robots. These can be bipedal, quadruped or hexapodic versions, or microscopic creations modeled on insects, such as crickets or grasshoppers, effectively to the point of producing what is a veritable robotic zoo.[3] These systems offer biologists and neuroscientists experimental training grounds for their models, and can potentially represent industrial tools to innovatively solve navigational or manipulation problems in sites difficult to access with a more traditional approach.

Perhaps on the skyscrapers in New York or Dubai we will see autonomous mini-robots, inspired by geckos, that adhere to the walls and clamber on up to clean the windows[4]—or we will be swallowing a robotic pill able to move inside our digestive system thanks to progression mechanisms inspired by inchworm locomotion to painlessly diagnose the stomach or colon.[5]

The emulation of what is natural is another face of bio-inspiration, and is very promising in engineering. This is to the extent of not stopping only at replication of the biomechanical and motor control part, but achieving the extreme complexity of entering into relation with the central nervous system, creating a scientific frontier area between neuroscience and robotics that takes the name of neuro-robotics.

One of the dreams of neuro-robotics is the connection between a patient's brain and a robotic system, with the aim of transmitting commands and receiving sensory feedback. The connection can come about thanks to a key component, the neural interface, consisting of a microdevice implanted in the nervous system in a surgical operation. Neural interfaces are a particular type of Brain Machine Interface, already defined in Chap. 4. By coming into direct contact with the peripheral or central nervous system in specific cerebral reference areas, they are able to read the signals of the neurons, and then stimulate the perception or interpret primary motor intentions. In practice, if the subject thinks of wanting to open and close the hand, the areas of the brain determining this movement, on which the electrodes are implanted, are activated. Consequently, the electrodes, recording the

---

[3]P. Dario, *Bioinspired Minirobotic Platforms for Educational Activities*, in *Advances in Autonomous Mini Robots*, Springer, 2012.

[4]K. Autumn et al., *Frictional adhesion: a new angle on gecko attachment*, «Journal of Experimental Biology», 209.18 (2006): 3569–3579.

[5]P. Dario, et al., *Development and in vitro testing of a miniature robotic system for computer-assisted colonoscopy*, Computer Aided Surgery 4.1 (1999): 1–14.

activation, interpret the thought, recognizing the intention specifically associated with moving the hand. In the opposite direction, if the interface is implanted in the areas of the central nervous system responsible for sensory feedback, the electrodes can stimulate tactile sensations as a result of a tactile signal obtained from an artificial sensor operating in the environment.

In the case of the 'peripheral' interfaces implanted in the peripheral nervous system[6]—for example at arm height, on the medial and ulnar nerve branching from the brachial plexus—the electrodes can connect both to afferent fibers, so inserting themselves into the circuit which carries the tactile signal from the fingertips to the brain, and to efferent fibers, so transmitting the motor intentions to the artificial hand. This establishes a bi-directional connection that can control grasping or evoke in the subject a tactile sensation similar to that which occurs when the mechanoreceptors in the skin of the natural hand enters into contact with the surrounding environment.

We can therefore say that neural interfaces are the bridge between the natural and the artificial, establishing a dialogue i.e. an exchange of signals, as studied in bionics.[7] Limit, bridge, dialogue: these are fundamental concepts of our journey through robotics, and of the relationship between the human and the artificial. The term bionics reminds the slightly more mature readers, who watched him on television, of the 'bionic man' depicted in the successful 1970s television series. It tells the story of a colonel in the US army who, after a war injury, loses his limbs and an eye, and undergoes the implantation of super prostheses thanks to which he achieves exceptional performances, *beyond the human.* This science fiction TV series has the merit of having foreseen, at least in part, what science later actually achieved, but also contributed to giving a magical or fictional aura to a sector that is in reality very serious and has achieved many important results.

Bionics models the principles of operation of natural systems through reproduction in artificial systems, but also studies implantable devices which,

---

[6]C. M. Oddo, et al. *Intraneural stimulation elicits discrimination of textural features by artificial fingertip in intact and amputee humans,* Elife 5 (2016): e09148.

[7]Bionics is defined as the "science that studies the sensorial and motor functions of living organisms (and their components), with the aim of reproducing them or enhancing them with devices of various kinds…". A year ago, there was the introduction in Pisa of the Master's degree in Bionics Engineering organised jointly by the University of Pisa and the Sant'Anna School of Advanced Studies. For my colleagues and myself, this was a very important act which consolidated the status of bionics as a scientific discipline, and as teaching subject with scientific dignity according to the Aristotelian interpretation. At the moment in which an engineering study program institutes a course with the title of a specific scientific area, it receives an academic approval that attributes both maturity and dignity. Due to the mere fact of being taught, we can say that also in Italy bionics is now accredited as a science and technology of the future, because within a year we shall have new engineers graduating in Bionics Engineering.

interfacing with living systems, interpret, enhance or amplify operational characteristics. The origin of modern bionics dates back to a scientific meeting that took place in 1960 in Ohio, USA. An exciting discussion was initiated on the emulation of natural characteristics with artificial systems, and on the treatment of information in biological systems, as reported in a famous article[8] published in Science, still very stimulating to read today.

In the bionic man TV series, technology enhances the performance of the man in which it is implanted, and we have super hearing or bionic sight more powerful than the human equivalents. However, scientific bionics was orig-inated to replace the operational features and restore senses such as hearing or sight in order to resolve a congenital or acquired disability. In clinical practice it has already obtained significant results in the field of profound deafness, and it is foreseen that important goals will be achieved in sight-related disabilities.

It should be noted that the bionic eye[9] is a system designed to restore the interrupted connection in a blind person between the central nervous system, in particular the optic nerve, and the retina. In fact it is called "bionic eye" because a neural interface is implanted on the structure of the eye which is able to read the external images through a camera and translate them into a series of coherent and synchronized electrical stimuli. These, once transmitted in the form of stimuli through the implanted electrodes to the optic nerve, evoke visual feedback and subsequently the vision of the object in the patient's brain. There are huge problems of biocompatibility at the interface between the artificial system and the natural one, particularly concerning those aspects related to the shape and material used for the mechanical construction of the electrodes and to the algorithms for the encoding of the information into electrical stimuli. In this sense, bionics can also be thought of as a system of interpretation and translation, from the language of neurons to that of external devices and vice versa. In order to understand its technical and scientific content, it is important to be very familiar with not only neurophysiology, but also the electronics and mechanics of the biomaterials needed to design the electrodes, and also the system that transcribes the external signal into electrical stimuli. While in the field of vision, research has yet to make significant progress, on the other hand it has already produced satisfactory results in the sector of hearing. In fact, a great success of bionics, perhaps the most important and the most clinically successful, is represented

[8]L. E. Lipetz, *Bionics*, Science, New Series, Vol. 133, No. 3452 (Feb. 24, 1961), pp. 588-590+592–593.
[9]Y. H-L., Luo, and L. da Cruz, *A review and update on the current status of retinal prostheses (bionic eye)*, British medical bulletin 109.1 (2014): 31–44.

by bionic hearing. An implantable prosthesis exists that communicates directly with the central nervous system, stimulating neurons as a consequence of sound inputs, and translating a series of sounds picked up by microphones into electrical impulses transmitted to the specific neurons by implantable microelectrodes, and this restores the ability to hear for the deeply deaf. The neuro-prostheses for hearing[10] can enable children to hear, and learn to interpret spoken language thanks to the implantable substitute system of natural hearing that overcomes deafness.

The interfaces based on electrodes that come into contact with the central nervous system are called neuro-prostheses,[11] and can be used for the senses, such as sight, hearing or touch, or aimed at recovering the movement of the upper or lower limbs in paralyzed subjects who have suffered a spinal injury. Since neurons in the central nervous system cannot be naturally regenerated, the only way to bypass the problem of breaking neural circuits is to use an interface that transmits signals to the robot in order to produce movement.

In the case of neuro-prostheses dedicated to motor control, they can be cortical implantable interfaces that read the signals emitted by the neurons of the brain areas where the motor signal associated with the intention to move is produced, and send commands to manipulators or external robots. This in turn generates a movement in line with the wishes of the subject who has undergone the implant.

Imagine a quadriplegic person[12] who cannot move their arms and legs and is totally dependent on external assistance. This subject is strongly motivated to reactivate a part of their autonomy, through the use of a system which, by expressing motor intention, enables control of the activation of a robotic system that can provide the desired action.

We have all seen videos available on the Net that show experiments carried out in specialized research centers in various parts of the world, which portray paralyzed patients with neuro-prosthesis implants, engaged in experiments in which they succeed for the first time independently to grasp a glass using a robotic arm.[13] It is evident the concentration and satisfaction of the patient,

[10]M. A. Svirsky et al. *Language development in profoundly deaf children with cochlear implants*, Psychological science 11.2 (2000): 153–158.

[11]D. Borton,, et al. *Personalized neuroprosthetics*, Science translational medicine 5.210 (2013): 210rv2–210rv2.

[12]J. A. Pruszynski , and J. Diedrichsen, *Reading the mind to move the body*, Science 348.6237 (2015): 860–861.

[13]To see an authoritative video example of a neural-implanted patient who recovers tactile feelings using a robotic arm, see http://www.theverge.com/2016/10/13/13269824/brain-implant-chip-feel-touch-robot-arm-paralyzed-tetraplegia.

who is strongly engaged in the experiment to provide and evaluate the brain-robot connection, which could not be tested in any other way.

Neuro-prostheses are made from the combination of electrodes that read signals, plus signal conditioning systems. These are coupled with software processing algorithms that allow reconnection of the expressed motor intention, which can no longer be transmitted naturally to the arm and hand due to spinal injury, to an external robotic system, which is operated as a result of the subject's command.[14] It is a kind of 'control through thought' made in an engineering way.

Only in cases of severe disability, such as quadriplegia, is an implant in the central nervous system acceptable, precisely because it is a very invasive intracranial operation, and it is necessary to approve the experimental protocol by ethical committees, which are very rigorous.

Now let us imagine a person who has had a hand amputated, and is eager to restore grip and manipulation. This is currently not guaranteed by the prostheses available on the market. These resemble more or less sophisticated pincers as part of a glove that resembles only aesthetically, but not functionally, the hand. The incentive for the creation of the artificial hand stems from the desire of amputees to completely recover their skills, and above all to feel and explore the external environment. For this reason, we researchers are working on the project of the bionic hand[15] connected with a neural interface implanted in the peripheral nervous system. This enables reconnection of hand and brain, and controls in a natural way the high performance level that can be achieved with advanced integrated robotics in a prosthetic hand.

The neural interfaces implanted in the peripheral nervous system do not interact with the primary motor intention in the brain, but are less invasive. They offer the advantage that with a single electrode implant you can connect both with the efferent fibers that transmit the motor command from the central nervous system up to the limb, and with the afferent fibers that follow the reverse path and stimulate the central nervous system by transferring the tactile sensory feedback from the hand to the brain. In the experiments we work to demonstrate the potential of peripheral interfaces that restore tactile sensation, thanks to stimuli recorded by the tactile sensors placed on the artificial fingertip, whose signals are then sent to the neuro-prosthesis itself to

---

[14]In the biomedical engineering sector in which these disciplines are studied, rehabilitation bioengineering, this situation is defined as support or functional substitution. The biomedical engineer studies and designs systems which exploit bionics and robotics in order to treat a specific problem and deals with the clinical implementation of the system.

[15]S. Raspopovic, et al. *Restoring natural sensory feedback in real-time bidirectional hand prostheses*, Science translational medicine 6.222 (2014): 222ra19–222ra19.

carry out the encoding, and coherently stimulate the fibers of the peripheral nervous system through the electrodes themselves.

Decoding refers to the reading of the neural signal and its translation into commands for external effectors, such as the hand prosthesis, while encoding is the translation of the signals captured by external sensors (cameras, tactile sensors or microphones depending on whether we are dealing with vision, touch or hearing) into electrical impulses to stimulate nerve fibers, and ultimately the brain that transforms sensory feedback into perception.

The hand prosthesis is not a robot, but its translation into a wearable medical device, which is implanted by means of a special socket on the stump of an amputee subject and enables the gripping and manipulating of objects in a manner similar to the natural hand.

Robotics offers methods and technologies for prosthetics to progress, and to obtain wearable devices that can substitute the functions of organs or parts of the human body.

I have worked for years on the artificial hand, investigating how to replicate touch, to obtain the maximum operational characteristics of the prostheses to restore the ability to grasp in activities of daily living and to 'feel' the characteristics of objects. This still today represents a challenge for amputees, for whom at the moment many movements with prostheses are still precluded.

Some examples of gestures of everyday life that we are accustomed to, but which with a prosthesis are very complex, if not impossible, are side grips. They are those actions like the one where you grasp a credit card or a thin object by taking it from the table. Another example is the strong grip that requires dexterity, useful for example to lace up shoes, or the grip required to button up a jacket, slotting the button into the buttonhole. These are movements that we do automatically, but are very difficult to achieve with a prosthetic hand because they require many levels of freedom and grip of the hand around the object. They are qualities that current prostheses, still in effect more or less evolved pincers, do not possess.

The natural hand is a sense organ and a versatile effector, equipped with dexterity that allows precision movements, which are very fine but which also require strong force, like that applied to lifting a heavy suitcase. It is perhaps one of the most fascinating and important parts of the human body to study and replicate with biomechanical models and with robotic systems that know how to imitate its movements.

Returning to the metaphor of the journey of robotics, in the scenario of hand prosthesis the robot is no longer *outside* of us, but *on* us, and it converses with the intimate part of us thanks to the neuro-prosthesis. This is how wearable robotics originates, which deals with the robots that we wear on our body and we carry with us, such as the *exoskeletons*.

To understand the purpose of an exoskeleton, let us proceed with an example connected to the use of bionics. In this context one of the aspects studied by research is a reduction in the invasiveness of neural interfaces. This is, for example, thanks to the realization of Brain Machine Interfaces that read the neural signals in a more indirect way. This avoids implants directly in the brain, as occurs with the external electrode pads, positioned on the skull for reading electroencephalographic signals.

In this area recently at the Sant'Anna School of Advanced Studies, my research group, together with other European groups, published a scientific article in *Science Robotics*.[16] It describes the experimental results of a project in which people with very high but incomplete spinal lesions, and who are able to control the movement of the approach of the arm to an object but not to open and close the hand, can complete the movement of the fingers thanks to an exoskeleton worn on the thumb and forefinger. The motor intention is read by a system of external non-invasive electrodes, and transmitted to the exoskeleton, which restores the grip on some objects, for some activities of daily life, for example drinking from a glass, taking cutlery for eating or holding a pen to write a signature.

In the scenario just described, the robot is *worn* on the subject's hand, like a semi-rigid dress that gently exerts a selective force on the fingers, to produce the different grips needed to grasp the objects. The exoskeleton is therefore worn to gradually become stronger and move better. If we have an intact, but very weak body, which we manage to control with difficulty, we can imagine a robotic suit, aesthetically attractive and silent in its movements, which reads motor intentions but also gently supports them, enhances them, and interacts appropriately with the environment.

To understand the potential of wearable robotics and its application perspectives, we can think of two examples of use. The first concerns a manufacturing environment and a work task, while the second regards the home and a typical daily chore.

In the first case, we can imagine a worker who has to perform a maintenance job in an industrial plant in a site that is difficult to reach. For this task he must stand in an uncomfortable position with his knees or body bent, and at the same time concentrate and stay still for a long time to complete the job. It is a matter of maintaining a demanding and arduous position over time,

---

[16]S. R. Soekadar, et al. *Hybrid EEG/EOG-based brain/neural hand exoskeleton restores fully independent daily living activities after quadriplegia*, Science Robotics 1.1 (2016): eaag3296.

which will probably cause back pain and other occupational pathologies that will not enable him to work beyond a certain age and at the same time causing discomfort and pain. How can robotics tackle this problem? Through the use of a semi-rigid and polyarticulated suit which, worn over his clothes, and active when necessary to provide him with the strength to support his position, compensating for gravity and thus giving him the sensation of "floating" while maintaining agility. Here the concept of wearability is combined with the softness, flexibility and lightness that are necessary to give a pleasant feeling of being assisted without being impeded in natural movements. In practice, we are presenting a progression of the concept of cognitive symbiosis to a more complete biomechanical symbiosis, which enables a better job to be done. In this scenario, in fact, it is not a matter of replacing a worker with a robot, but of developing a robot worn by the worker to function better and use his skills and competences without being exposed to the typical wear and tear of his work, which in the long term may cause problems in the musculoskeletal system.

The second scenario concerns the house, where a person paralyzed by spinal injury can no longer control the movement of his intact lower limbs and due to this he is confined to a wheelchair. Imagine that, at least for a few hours a day, our subject, with simple maneuvers, can wear an exoskeleton that can give them the strength necessary to activate their skeletal system, stand up and walk out of the house. The exoskeleton therefore manages to bypass the interruption in the central nervous system following the accident and to provide the power and strength sufficient to support walking and balance. It is clear that also in this second case it is very important to obtain the human-machine symbiosis which transmits motor intention from the brain to the exoskeleton, and the return of sensorial feedback from the exoskeleton sensors, enabling perception and therefore full motor control.

We have therefore described two profiles of people who are very different in their characteristics, but identical in terms of the need for assistance, i.e. they can receive help from the wearable robotic system to better perform their motor tasks.

Today there are more than six hundred million elderly people in the world and we know from the demographic forecasts that this number should double from now to 2030. If we extend to older users the use of exoskeletons, we can imagine that they become like wearable overalls, which can give strength, resistance and assistance in movement, enough to overcome the limits imposed by aging, fatigue, or debilitating diseases. Exoskeletons can solve one of the main problems resulting from aging, that of the personal assistance increasingly necessary in the last part of life to keep the 'journey' autonomous.

Wearable robotics embodies and extends the concept of human machine symbiosis, from the extension of intelligence[17] to an extension of strength, obtained through exoskeletons to support the movement of limbs, for rehabilitation purposes or for personal assistance.

The term *exoskeleton* is bio-inspired because it refers to biological organisms that have a semi-rigid skeleton outside the body, such as crustaceans, as opposed to what there is in the human species, which is equipped with an endoskeleton. There are very recent and innovative ideas that offer alternative visions with active suits, but in general in robotics the exoskeleton has rigid segments that cover at least part of the limbs and joints. These, with axes parallel to the axes of our joints, support natural movement by amplifying it, by means of an additional torque, and in fact representing an artificial musculoskeletal system activated in parallel with the natural one. It is therefore a mechanical amplifier of motor intention, which strengthens the weak one expressed by the body, giving support, speed and amplitude in the person's movements.

The two families of exoskeletons (upper limbs and lower limbs) are very distinct from each other. This is not so much in terms of mechanical construction, which however follows similar principles appropriately adapted to different joints, but they are dissimilar especially in the interface and control mechanism, because motor control is very different from that of gripping and manipulation. The exoskeletons for the upper limb usually involve the shoulder, the elbow, the wrist and the hand with the fingers, and are designed to perform all movements and degrees of freedom of the upper limb. This is so as to be able to guide the movements of the upper limb of the subject wearing the exoskeleton, with different control modalities. One is active, which strengthens the weak movement initiated by the subject by providing an additional torque, and the other instead exercises a guide to the movement of the passive limb that lets the exoskeleton 'gently drag' where the subject wishes to go. In a very similar way, but with a different control scheme because the control of the walking is automatic and oscillatory, the lower limb exoskeletons activate the joints of the hip, knee and ankle, in order to support the walking of weak or paralyzed subjects. Due to spinal injuries they are unable to control the intact musculoskeletal system.

---

[17]Recently a document was issued by the Executive Office of President Obama, edited by the National Science and Technology Council Committee on Technologies, *Preparing for the Future of Artificial Intelligence*, October 2016 https://obamawhitehouse.archives.gov/sites/default/files/whitehouse_files/microsites/ostp/NSTC/preparing_for_the_future_of_ai.pdf.

For an elderly or weak subject, wearing a lower limb exoskeleton, which wraps around the legs and pushes them when necessary, it is like riding an electric bicycle, which when on the flat does not resist and allows the subject's pedaling to drive the movement. However, when it goes uphill, it adds a torque to the pedal stroke thanks to the electric motor, and gives an extra push necessary so as not to tire the rider excessively. The ideal functioning of the wearable robot is similar. There is docility or transparency with respect to the movement that can be expressed in a natural way, and the push imparted to achieve the desired movement when there is not enough force to carry it out.

There are two other possible uses of exoskeletons in the medical field: rehabilitation and personal assistance. The first concerns the administration of rehabilitative therapies necessary after an orthopedic trauma or after a cardiovascular disease such as stroke, which both leave consequences in terms of hemiparesis, to be treated with appropriate exercises. In this case the robot is used for exercises which, by exploiting the plasticity of the central nervous system and the muscular mechanisms and synergies that are reactivated with the movement, facilitate the training and re-learning of the motor patterns that will be consolidated over time. In the second case, the wearable robot supports the movement as personal assistant, not so much for rehabilitation purposes, but as a support to the movement.

Both for rehabilitation and for personal assistance, it is essential to achieve a symbiosis between the exoskeleton and the person who is wearing it. Through a training phase the user must learn to manage a different motor and operating scheme and, by exploiting the strength of the robot, learn to overcome weakness and carry out motor action.

In reality, in the United States, exoskeletons were born for military applications, with the idea that they could be used for the physical augmentation of soldiers. This is in order to be able to bear greater loads safely and for a longer time, and to endure and walk faster. The application to military enhancement did not work very well for various reasons, which we do not have the time to go into here. However, it did lead to the development of technologies and know-how that are being transferred to the medical sectors, to disabled people, and to industrial automation for the functional support of subjects who perform strenuous and tiring physical activities and consequently are exposed to the risk of occupational pathologies of the musculoskeletal system.

Recently, in Europe and in Asia, together with the United States, there has been great interest from large industrial players in exoskeletons. There is a veritable international competition under way in order to develop better devices in order to put on the market[18] innovative products for various civil and medical applications.[19]

The long-term dream of neuro-robotics is to give back to paralyzed people the ability to move and feel, and regain the operational characteristics of their body. This scientific work has been progressing for many years, and the comforting results that have been achieved in all sectors, from hearing, to sight, to the restoration of hand movement and the first very simple tactile sensations, make it possible to hope that we can obtain still greater progress.

In conclusion, we can say that neuro-robotics, studies, models, and interfaces natural with artificial, external and implanted, and gradually leads to creating man-robot symbiosis and to developing the bionic man, in which the functionality of damaged body parts, due to traumas or diseases, is restored through technology. From the point of view of neuro-robotics, the study of the connection of the robot with the brain is a test bed on which the quality of engineering solutions and the validity of technologies are measured.

The progression of the artificial hand tangibly illustrates the evolution that we have described in this book. First of all, industrial robotics was also originated to develop hands, or pincers, able to support or replace workers' hands on the assembly line, and the first examples of robots were robotic arms with clamps, which performed repetitive tasks at high speed and with reliability. Even today, a benchmarking criterion of industrial automation industries is represented precisely by the quality of the robotic arms they produce, by their lightness and dexterity. An emblematic case is that of the new sector of collaborative robotics, in which the objective is precisely that of combining the robotic arm with the human operator in a safe way, controlling the interaction with the person, or the environment, in order to collaborate and enhance the movement. In collaborative robotics we want to manage the movement of the robot in contact with a worker thanks to sophisticated interaction control algorithms implemented in high performance robotic arms and equipped with a new generation of sensor matrices analogous to an artificial skin covering the robot body itself.

As a completion of the arm, in order to make it functional, the robotic hand represents an important challenge for industrial robotics, and especially for service robotics. In fact, creating the hand represents the effector of the

---

[18] Also the author is founding partner of IUVO, a spinoff of the *Scuola Superiore Sant'Anna*, which deals with wearable robotics and exoskeletons.

[19] In the European Union, the framework programs have funded research for medical applications, supporting the research into developing rehabilitation or personal assistance exoskeletons, and within a few years results are expected from industrial exploitation of the intellectual property generated.

logistics robot, enabling it to take and transport objects, or to teleoperate the surgical procedure by replacing the surgeon's hand at the operation site. Finally, the artificial hand can prompt a leap in quality for the social robot, which offers us a service by carrying out actions for the preparation of food, or for other domestic tasks. Without the ability to reach, take, grasp and manipulate objects, we will not have a social robot able to truly enter the service industry. We can say that the robot can transport objects but cannot carry out the initial and final parts of the task, which consists in the manipulation of the object to be carried.

This series of examples is useful to understand the concept of robotics that is gradually leaving the factories, entering hospitals, warehouses, and finally coming into our homes and onto our streets. But the metaphor that really closes the circle and makes us understand the potential of the industrial revolution we are going through is that of robotics "within us". This is the robotics which, through bionics and neuroprosthetics, manages not only to implant itself in our body, but also to communicate intimately with our brain, achieving that intuition of Licklider which envisioned a computer-user symbiosis. The difference is that we are not just talking about a computer but about a robot, which has the ability to generate force and movement.

As a progressive, I am convinced that we must move forward in the study of robotics which, integrated with artificial intelligence, information technology and communication technologies, will enable us to overcome many limitations. The technologies described in this book may very well upset our model of society, transform or eliminate jobs, and enter into contact with our cognitive intimacy. So more in-depth studies are required from every point of view, due to the implications that can emerge, and the social and human consequences. However, I believe that a philosophical and humanistic effort is also crucial to accompany these advances and help us to interpret their impact on society and on humanity, with reflections in the legal field of law and rights, and the ethical and civil responsibility of those who develop or use these systems. Being progressive, in the midst of this era of the robotic revolution, means acquiring awareness through study, the diffusion of culture, and the open discussion of these issues.

At the threshold of the fourth revolution, we have before us an important challenge of training and antidisciplinary research for the school system and for the university, If seized in full, this can prompt a transition of our culture with complete respect for the dignity of the person. If some trades and jobs are canceled by this fourth industrial revolution, it is likely that others will be created. But if this happens to a lesser or greater extent will also depend on our ability to understand and foresee things, and on our ability to go beyond limits and then to find new ones within the cycle of scientific research, or to build bridges to extend our intelligence. As I have been Minister of Education, I hope that Italy and European Union, choose to be protagonist in this transition, and that is why investments are needed, in education, in research and in advanced training.

Acknowledgements I wish to thank all my family, and in particular my children Andrea and Marianna, for having been close to me in the ups and downs of my recent life, with their affection, support and understanding.

There are some people to whom I feel profound gratitude: my colleagues and students of the Sant'Anna School of Advanced Studies with whom I have collaborated and shared many projects over the years, especially my teacher Paolo Dario, who taught me the fundamentals of research and fought for the academic and scientific future of all of us, and my students Calogero, Christian and Nicola, because they make me feel proud of them.

I would like to say a special thank you to Alessandro Aresu, my 'Editor', for the firm friendship and professionalism with which he constantly stimulated my writing. Thank you also to Marco Meloni and to the whole School of Politics for having cultivated the idea of this book.

The rereading of the manuscript was carried out with great patience and rigor by Luigi Bimbi, Lia Marianelli, Nicola Vitiello and Calogero Oddo. Each of them dedicated time and great skill in suggesting how to improve the work.

Printed in the United States
By Bookmasters